Contents

Structural Vibration: Analysis and Damping

C. F. Beards BSc, PhD, C Eng, MRAeS, MIOA

Consultant in Dynamics, Noise and Vibration
Formerly of Imperial College of Science,
Technology and Medicine,
University of London

ARNOLD

A member of the Hodder Headline Group
LONDON • SYDNEY • AUCKLAND

Copublished in the Americas by Halsted Press
an imprint of John Wiley & Sons Inc.
New York – Toronto

First published in Great Britain 1996 by Arnold,
a member of the Hodder Headline Group,
338 Euston Road, London NW1 3BH

Copublished in the Americas by Halsted Press,
an imprint of John Wiley & Sons Inc.,
605 Third Avenue,
New York, NY 10158–0012

British Library Cataloguing in Publication Data
A catalogue record for this book is available from the British Library

Library of Congress Cataloging-in-Publication Data
A catalog record for this book is available from the Library of Congress

ISBN 0 340 64580 6
ISBN 0 470 23586 1 (Wiley only)

Typeset in 10/12 Times by Poole Typesetting (Wessex) Ltd, Bournemouth
Printed and bound in Great Britain by J W Arrowsmith Ltd, Bristol

Preface

The analysis of structural vibration is necessary in order to calculate the natural frequencies of a structure, and the response to the expected excitation. In this way it can be determined whether a particular structure will fulfil its intended function and, in addition, the results of the dynamic loadings acting on a structure can be predicted, such as the dynamic stresses, fatigue life and noise levels. Hence the integrity and usefulness of a structure can be maximized and maintained. From the analysis it can be seen which structural parameters most affect the dynamic response so that if an improvement or change in the response is required, the structure can be modified in the most economic and appropriate way. Very often the dynamic response can only be effectively controlled by changing the damping in the structure. There are many sources of damping in structures to consider and the ways of changing the damping using both active and passive methods require an understanding of their mechanism and control. For this reason a major part of the book is devoted to the damping of structural vibrations.

Structural Vibration: Analysis and Damping benefits from my earlier book *Structural Vibration Analysis: Modelling, Analysis and Damping of Vibrating Structures* which was published in 1983 but is now out of print. This enhanced successor is far more comprehensive with more analytical discussion, further consideration of damping sources and a greater range of examples and problems. The mathematical modelling and vibration analysis of structures are discussed in some detail, together with the relevant theory. It also provides an introduction to some of the excellent advanced specialized texts that are available on the vibration of dynamic systems. In addition, it describes how structural parameters can be changed to achieve the desired dynamic performance and, most importantly, the mechanisms and methods for controlling structural damping.

It is intended to give engineers, designers and students of engineering to first degree level a thorough understanding of the principles involved in the analysis of structural vibration and to provide a sound theoretical basis for further study.

There is a large number of worked examples throughout the text, to amplify and clarify the theoretical analyses presented, and the meaning and interpretation of the results obtained are fully discussed. A comprehensive range of problems has been included, together with many worked solutions which considerably enhance the range, scope and usefulness of the book.

Chris Beards
August 1995

Acknowledgements

Some of the problems first appeared in University of London B.Sc. (Eng) Degree Examinations, set for students of Imperial College, London. The section on random vibration has been reproduced with permission from the *Mechanical Engineers Reference Book*, 12th edn (Butterworth–Heinemann, 1993).

*If a structure is built for a man and the builder
does not make its design and construction meet
the requirements and it collapses in whole or in
part, then the builder shall strengthen and
restore the structure at his own expense.*

*If other property is damaged or destroyed by the
collapse then the builder shall restore that also
at his own expense.*

*If the collapse causes loss of life then the
builder shall be put to death.*

The Code of Hammurabi, *c.* 1750 BC

General notation

a	damping factor, dimension, displacement.
b	circular frequency (rad/s), dimension.
c	coefficient of viscous damping, velocity of propagation of stress wave.
c_c	coefficient of critical viscous damping $= 2\sqrt{(mk)}$.
c_d	equivalent viscous damping coefficient for dry friction damping $= 4F_d/\pi\omega X$.
c_H	equivalent viscous damping coefficient for hysteretic damping $= \eta k/\omega$.
d	diameter.
f	frequency (Hz), exciting force.
f_s	Strouhal frequency (Hz).
g	acceleration constant.
h	height, thickness.
j	$\sqrt{(-1)}$.
k	linear spring stiffness, beam shear constant.
k_T	torsional spring stiffness.
$k*$	complex stiffness $= k(1 + j\eta)$.
l	length.
m	mass.

q	generalized coordinate.
r	radius.
s	Laplace operator $= a + jb$.
t	time.
u	displacement.
v	velocity,
	deflection.
x	displacement.
y	displacement.
z	displacement.
A	amplitude,
	constant,
	cross-sectional area.
B	constant.
$C_{1,2,3,4}$	constants.
C_D	drag coefficient.
D	flexural rigidity $= Eh^3/12(1 - v^2)$,
	hydraulic mean diameter.
D	derivative w.r.t. time.
E	modulus of elasticity.
E'	in-phase, or storage modulus.
E''	quadrature, or loss modulus.
$E*$	complex modulus $= E' + jE''$.
F	exciting force amplitude.
F_d	Coulomb (dry) friction force (μN).
F_T	transmitted force.
G	centre of mass,
	modulus of rigidity.
I	mass moment of inertia.
J	second moment of area,
	moment of inertia.
K	stiffness,
	gain factor.
L	length.
M	mass,
	moment,
	mobility.
N	applied normal force,
	gear ratio.
P	force.
Q	factor of damping.
Q_i	generalized external force.
R	radius of curvature.
S	Strouhal number,
	vibration intensity.

$[S]$	system matrix.
T	kinetic energy,
	tension,
	time constant.
T_{R}	transmissibility $= F_{\mathrm{T}}/F.$
V	potential energy,
	speed.
X	amplitude of motion.
$\{X\}$	column matrix.
X_{S}	static deflection $= F/k$, where k is linear stiffness.
X/X_{S}	dynamic magnification factor.
Z	impedance,
	vibration intensity.
α	coefficient,
	influence coefficient,
	phase angle,
	receptance.
β	coefficient,
	receptance.
γ	coefficient,
	receptance.
δ	deflection.
ε	short time,
	strain.
ε_0	strain amplitude.
ζ	damping ratio $= c/c_{\mathrm{c}}.$
η	loss factor $= E''/E'.$
θ	angular displacement,
	slope.
λ	matrix eigenvalue,
	$[\rho A\omega^2/EI]^{1/4}.$
μ	coefficient of friction,
	mass ratio $= m/M.$
ν	viscosity,
	Poisson's ratio,
	circular exciting frequency (rad/s).
ξ	time.
ρ	material density.
σ	stress.
σ_0	stress amplitude.
τ	period of vibration $= 1/f.$
τ_{d}	period of dry friction damped vibration.
τ_{v}	period of viscous damped vibration.
ϕ	phase angle,
	function of time,
	angular displacement.

ψ phase angle.
ω undamped circular frequency (rad/s).
ω_d dry friction damped circular frequency.
ω_v viscous damped circular frequency $= \omega\sqrt{(1 - \zeta^2)}$.
Λ logarithmic decrement $= \ln X_1/X_{11}$.
Ω natural circular frequency (rad/s).

1

Introduction

A structure is a combination of parts fastened together to create a supporting framework, which may be part of a building, ship, machine, space vehicle, engine or some other system.

Before the Industrial Revolution started, structures usually had a very large mass because heavy timbers, castings and stonework were used in their fabrication; also the vibration excitation sources were small in magnitude so that the dynamic response of structures was extremely low. Furthermore, these constructional methods usually produced a structure with very high inherent damping, which also gave a low structural response to dynamic excitation. Over the last 200 years, with the advent of relatively strong lightweight materials such as cast iron, steel and aluminium, and increased knowledge of the material properties and structural loading, the mass of structures built to fulfil a particular function has decreased. The efficiency of engines has improved and, with higher rotational speeds, the magnitude of the vibration exciting forces has increased. This process of increasing excitation with reducing structural mass and damping has continued at an increasing pace to the present day when few, if any, structures can be designed without carrying out the necessary vibration analysis, if their dynamic performance is to be acceptable.

The vibration that occurs in most machines, structures and dynamic systems is undesirable, not only because of the resulting unpleasant motions, the noise and the dynamic stresses which may lead to fatigue and failure of the structure or machine, but also because of the energy losses and the reduction in performance that accompany the vibrations. It is therefore essential to carry out a vibration analysis of any proposed structure.

There have been very many cases of systems failing or not meeting performance targets because of resonance, fatigue or excessive vibration of one component or another.

Because of the very serious effects that unwanted vibrations can have on dynamic systems, it is essential that vibration analysis be carried out as an inherent part of their design; when necessary modifications can most easily be made to eliminate vibration or at least to reduce it as much as possible.

It is usually much easier to analyse and modify a structure at the design stage than it is to modify a structure with undesirable vibration characteristics after it has been built. However, it is sometimes necessary to be able to reduce the vibration of existing structures brought about by inadequate initial design, by changing the function of the structure or by changing the environmental conditions, and therefore techniques for the analysis of structural vibration should be applicable to existing structures as well as to those in the design stage. It is the solution to vibration problems that may be different depending on whether or not the structure exists.

To summarize, present-day structures often contain high-energy sources which create intense vibration excitation problems, and modern construction methods result in structures with low mass and low inherent damping. Therefore careful design and analysis is necessary to avoid resonance or an undesirable dynamic performance.

1.1   THE CAUSES AND EFFECTS OF STRUCTURAL VIBRATION

There are two factors that control the amplitude and frequency of vibration in a structure: the excitation applied and the response of the structure to that particular excitation. Changing either the excitation or the dynamic characteristics of the structure will change the vibration stimulated.

The excitation arises from external sources such as ground or foundation vibration, cross winds, waves and currents, earthquakes and sources internal to the structure such as moving loads and rotating or reciprocating engines and machinery. These excitation forces and motions can be periodic or harmonic in time, due to shock or impulse loadings, or even random in nature.

The response of the structure to excitation depends upon the method of application and the location of the exciting force or motion, and the dynamic characteristics of the structure such as its natural frequencies and inherent damping level.

In some structures, such as vibratory conveyors and compactors, vibration is encouraged, but these are special cases and in most structures vibration is undesirable. This is because vibration creates dynamic stresses and strains which can cause fatigue and failure of the structure, fretting corrosion between contacting elements and noise in the environment; also it can impair the function and life of the structure or its components (see Fig. 1.1).

1.2   THE REDUCTION OF STRUCTURAL VIBRATION

The level of vibration in a structure can be attenuated by reducing either the excitation, or the response of the structure to that excitation or both (see Fig. 1.2). It is sometimes possible, at the design stage, to reduce the exciting force or motion by changing the equipment responsible, by relocating it within the structure or by isolating it from the structure so that the generated vibration is not transmitted to the supports. The structural response can be altered by changing the mass or stiffness of the structure, by moving the

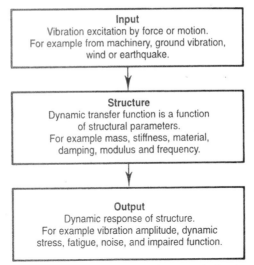

Fig. 1.1.   Causes and effects of structural vibration.

source of excitation to another location, or by increasing the damping in the structure. Naturally, careful analysis is necessary to predict all the effects of any such changes, whether at the design stage or as a modification to an existing structure.

Suppose, for example, it is required to increase the natural frequency of a simple system by a factor of two. It is shown in Chapter 2 that the natural frequency of a body of mass m supported by a spring of stiffness k is $(1/2\pi) . \sqrt{(k/m)}$ Hz, so that a doubling of this

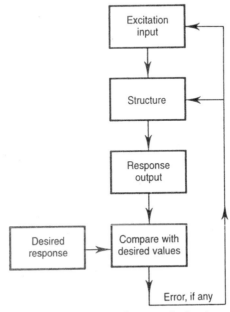

Fig. 1.2.   Reduction of structural vibration.

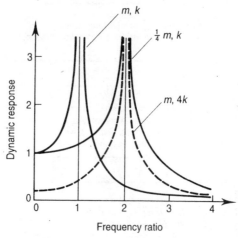

Fig. 1.3.   Effect of mass and stiffness changes on dynamic response.

frequency can be achieved either by reducing m to $\frac{1}{4}m$ or by increasing k to $4k$. The effect of these changes on the dynamic response is shown in Fig. 1.3. Whilst both changes have the desired effect on the natural frequency, it is clear that the dynamic responses at other frequencies are very different.

The Dynamic Transfer Function (DTF) becomes very large and unwieldly for complicated structures, particularly if all damping sources and non-linearities are included. It may be that at some time in the future all structural vibration problems will be solved by a computer program that uses a comprehensive DTF (Fig. 1.4). At present, however, analysis techniques usually limit the scope and hence the size of the DTF in some way such as by considering a restricted frequency range or by neglecting damping or non-linearities. Structural vibration research is currently aimed at a large range of problems from bridge and vehicle vibration through to refined damping techniques and measurement methods.

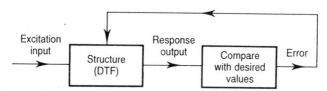

Fig. 1.4.   Feedback to modify structure to achieve desired levels.

1.3   THE ANALYSIS OF STRUCTURAL VIBRATION

It is necessary to analyse the vibration of structures in order to predict the natural frequencies and the response to the expected excitation. The natural frequencies of the structure must be found because if the structure is excited at one of these frequencies resonance occurs, with resulting high vibration amplitudes, dynamic stresses and noise levels. Accordingly resonance should be avoided and the structure designed so that it is not encountered during normal conditions; this often means that the structure need only be analysed over the expected frequency range of excitation.

Although it may be possible to analyse the complete structure, this often leads to a very complicated analysis and the production of much unwanted information. A simplified mathematical model of the structure is therefore usually sought that will, when analysed, produce the desired information as economically as possible and with acceptable accuracy. The derivation of a simple mathematical model to represent the dynamics of a real structure is not easy, if the model is to produce useful and realistic information. It is often desirable for the model to predict the location of nodes in the structure. These are points of zero vibration amplitude and are thus useful locations for the siting of particularly delicate equipment. Also, a particular mode of vibration cannot be excited by forces applied at one of its nodes.

Vibration analysis can be carried out most conveniently by adopting the following three-stage approach:

Stage   I. Devise a mathematical or physical model of the structure to be analysed.
Stage II. From the model, write the equations of motion.
Stage III. Evaluate the structure response to a relevant specific excitation.

These stages will now be discussed in greater detail.

1.3.1 Stage I. The mathematical model

Although it may be possible to analyse the complete dynamic structure being considered, this often leads to a very complicated analysis, and the production of much unwanted information. A simplified mathematical model of the structure is therefore usually sought that will, when analysed, produce the desired information as economically as possible and with acceptable accuracy. The derivation of a simple mathematical model to represent the dynamics of a real structure is not easy, if the model is to give useful and realistic information.

All real structures possess an infinite number of degrees of freedom; that is, an infinite number of coordinates are necessary to specify completely the position of the structure at any instant of time. A structure possesses as many natural frequencies as it has degrees of freedom, and if excited at any of these natural frequencies a state of resonance exists, so that a large amplitude vibration response occurs. For each natural frequency the structure has a particular way of vibrating so that it has a characteristic shape, or mode of vibration, at each natural frequency.

Fortunately it is not usually necessary to calculate all the natural frequencies of a structure; this is because many of these frequencies will not be excited and in any case they may give small resonance amplitudes because the damping is high for that particular mode of vibration. Therefore, the analytical model of a dynamic structure need have only a few degrees of freedom, or even only one, provided the structural parameters are chosen so that the correct mode of vibration is modelled. It is never easy to derive a realistic and useful mathematical model of a structure, because the analysis of particular modes of vibration is usually sought, and the determination of the relevant structural motions and parameters for the mathematical model needs great care.

However, to model any real structure a number of simplifying assumptions can often be made. For example, a distributed mass may be considered as a lumped mass, or the effect of damping in the structure may be ignored, particularly if only resonance frequencies are needed or the dynamic response required at frequencies well away from a resonance. A non-linear spring may be considered linear over a limited range of extension, or certain elements and forces may be ignored completely if their effect is likely to be small. Furthermore, the directions of motion of the mass elements are usually restrained to those of immediate interest to the analyst.

Thus the model is usually a compromise between a simple representation that is easy to analyse but may not be very accurate, and a complicated but more realistic model which is difficult to analyse but gives more useful results. Some examples of models derived for real structures are given below, whilst further examples are given throughout the text.

The swaying oscillation of a chimney or tower can be investigated by means of a single degree of freedom model. This model would consider the chimney to be a rigid body resting on an elastic soil. To consider bending vibration in the chimney itself would require a more refined model such as the four degree of freedom system shown in Fig. 1.5. By giving suitable values to the mass and stiffness parameters a good approximation to the first bending mode frequency, and the corresponding mode shape, may be obtained. Such a model would not be sufficiently accurate for predicting the frequencies of higher modes; to accomplish this a more refined model with more mass elements and therefore more degrees of freedom would be necessary.

Vibrations of a machine tool can be analysed by modelling the machine structure by the two degree of freedom system shown in Fig. 1.6. In the simplest analysis the bed can be considered to be a rigid body with mass and inertia, and the headstock and tailstock are each modelled by lumped masses. The bed is supported by springs at each end as shown. Such a model would be useful for determining the lowest or fundamental natural frequency of vibration. A refinement to this model, which may be essential in some designs of machine where the bed cannot be considered rigid, is to consider the bed to be a flexible beam with lumped masses attached as before.

To analyse the torsional vibration of a radio telescope when in the vertical position a five degree of freedom model, as shown in Fig. 1.7, can be used. The mass and inertia of

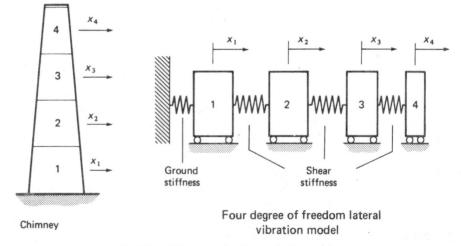

Chimney

Four degree of freedom lateral
vibration model

Fig. 1.5.   Chimney vibration analysis model.

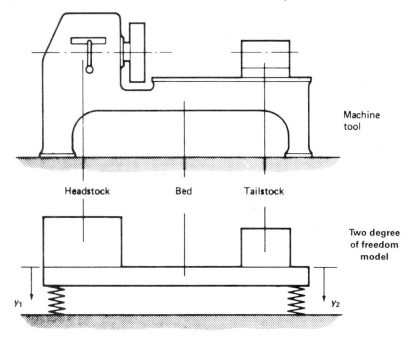

Fig. 1.6.   Machine tool vibration analysis model.

the various components may usually be estimated fairly accurately, but calculation of the stiffness parameters at the design stage may be difficult; fortunately the natural frequencies are proportional to the square root of the stiffness. If the structure, or a similar one, is already built, the stiffness parameters can be measured. A further simplification of the model would be to put the turret inertia equal to zero, so that a three degree of freedom model is obtained. Such a model would be easy to analyse and would predict the lowest natural frequency of torsional vibration with fair accuracy, provided the correct inertia and stiffness parameters were used. It could not be used for predicting any other modes of vibration because of the coarseness of the model. However, in many structures only the lowest natural frequency is required, since if the structure can survive this frequency it will be able to survive other natural frequencies too.

None of these models include the effect of damping in the structure. Damping in most structures is very low so that the difference between the undamped and the damped natural frequencies is negligible. It is only necessary to include the effect of damping in the model if the response to a specific excitation is sought, particularly at frequencies in the region of a resonance.

1.3.1.1 The model parameters

Because of the approximate nature of most models, whereby small effects are neglected and the environment is made independent of the system motions, it is usually reasonable to assume constant parameters and linear relationships. This means that the coefficients in the equations of motion are constant and the equations themselves are linear: these are real

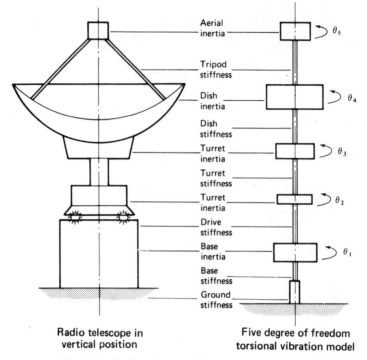

Aerial inertia — θ_5

Tripod stiffness

Dish inertia — θ_4

Dish stiffness

Turret inertia — θ_3

Turret stiffness

Turret inertia — θ_2

Drive stiffness

Base inertia — θ_1

Base stiffness

Ground stiffness

Radio telescope in
vertical position

Five degree of freedom
torsional vibration model

Fig. 1.7.   Radio telescope vibration analysis model.

aids to simplifying the analysis. Distributed masses can often be replaced by lumped mass elements to give ordinary rather than partial differential equations of motion. Usually the numerical value of the parameters can, substantially, be obtained directly, from the system being analysed. However, model system parameters are sometimes difficult to assess, and then an intuitive estimate is required, engineering judgement being of the essence.

It is not easy to create a relevant mathematical model of the structure to be analysed, but such a model does have to be produced before Stage II of the analysis can be started. Most of the material in subsequent chapters is presented to make the reader competent to carry out the analyses described in Stages II and III. A full understanding of these methods will be found to be of great help in formulating the mathematical model referred to above in Stage I.

1.3.2 Stage II. The equations of motion

Several methods are available for obtaining the equations of motion from the mathematical model, the choice of method often depending upon the particular model and personal preference. For example, analysis of the free-body diagrams drawn for each body of the model usually produces the equations of motion quickly, but it can be advantageous in some cases to use an energy method such as the Lagrange equation.

From the equations of motion the characteristic or frequency equation is obtained, yielding data on the natural frequencies, modes of vibration, general response and stability.

1.3.3 Stage III. Response to specific excitation

Although Stage II of the analysis gives much useful information on natural frequencies, response and stability, it does not give the actual response of the structure to specific excitations. It is necessary to know the actual response in order to determine such quantities as dynamic stress or noise for a range of inputs, either force or motion, including harmonic, step and ramp. This is achieved by solving the equations of motion with the excitation function present.

Remember:

> Stage   I.   Model
> Stage   II.   Equations
> Stage   III.   Excitation

1.4   OUTLINE OF THE TEXT

A few examples have been given above to show how real structures can be modelled, and the principles of their analysis. To be competent to analyse these models it is first necessary to study the analysis of damped and undamped, free and forced vibration of single degree of freedom structures such as those discussed in Chapter 2. This not only allows the analysis of a wide range of problems to be carried out, but is also essential background to the analysis of structures with more than one degree of freedom, which is considered in Chapter 3. Structures with distributed mass, such as beams and plates, are analysed in Chapter 4.

The damping that occurs in structures and its effect on structural response is described in Chapter 5, together with measurement and analysis techniques for damped structures, and methods for increasing the damping in structures. Techniques for reducing the excitation of vibration are also discussed. These chapters contain a number of worked examples to aid the understanding of the techniques described, and to demonstrate the range of application of the theory.

Methods of modelling and analysis, including computer methods of solution are presented without becoming embroiled in computational detail. It must be stressed that the principles and analysis methods of any computer program used should be thoroughly understood before applying it to a vibration problem. Round-off errors and other approximations may invalidate the results for the structure being analysed.

Chapter 6 is devoted entirely to a comprehensive range of problems to reinforce and expand the scope of the analysis methods. Chapter 7 presents the worked solutions and answers to many of the problems contained in Chapter 6. There is also a useful bibliography and index.

2

The vibration of structures with one degree of freedom

All real structures consist of an infinite number of elastically connected mass elements and therefore have an infinite number of degrees of freedom; hence an infinite number of coordinates are needed to describe their motion. This leads to elaborate equations of motion and lengthy analyses. However, the motion of a structure is often such that only a few coordinates are necessary to describe its motion. This is because the displacements of the other coordinates are restrained or not excited, being so small that they can be neglected. Now, the analysis of a structure with a few degrees of freedom is generally easier to carry out than the analysis of a structure with many degrees of freedom, and therefore only a simple mathematical model of a structure is desirable from an analysis viewpoint. Although the amount of information that a simple model can yield is limited, if it is sufficient then the simple model is adequate for the analysis. Often a compromise has to be reached, between a comprehensive and elaborate multi-degree of freedom model of a structure which is difficult and costly to analyse but yields much detailed and accurate information, and a simple few degrees of freedom model that is easy and cheap to analyse but yields less information. However, adequate information about the vibration of a structure can often be gained by analysing a simple model, at least in the first instance.

The vibration of some structures can be analysed by considering them as a one degree or single degree of freedom system; that is, a system where only one coordinate is necessary to describe the motion. Other motions may occur, but they are assumed to be negligible compared with the coordinate considered.

A system with one degree of freedom is the simplest case to analyse because only one coordinate is necessary to describe the motion of the system completely. Some real systems can be modelled in this way, either because the excitation of the system is such that the vibration can be described by one coordinate, although the system could vibrate in other directions if so excited, or the system really is simple as, for example, a clock pendulum. It should also be noted that a one, or single degree of freedom model of a complicated system can often be constructed where the analysis of a particular mode of vibration is to be carried out. To be able to analyse one degree of freedom systems is therefore essential in the analysis of structural vibrations. Examples of structures and motions which can be analysed by a single degree of freedom model are the swaying of a tall rigid building resting on an elastic soil, and the transverse vibration of a bridge. Before considering these examples in more detail, it is necessary to review the analysis of vibration of single degree of freedom dynamic systems. For a more comprehensive study see *Engineering Vibration Analysis with Application to Control Systems* by C. F. Beards (Edward Arnold, 1995). It should be noted that many of the techniques developed in single degree of freedom analysis are applicable to more complicated systems.

2.1   FREE UNDAMPED VIBRATION

2.1.1 Translation vibration

In the system shown in Fig. 2.1 a body of mass m is free to move along a fixed horizontal surface. A spring of constant stiffness k which is fixed at one end is attached at the other end to the body. Displacing the body to the right (say) from the equilibrium position causes a spring force to the left (a restoring force). Upon release this force gives the body an acceleration to the left. When the body reaches its equilibrium position the spring force is zero, but the body has a velocity which carries it further to the left although it is retarded by the spring force which now acts to the right. When the body is arrested by the spring the spring force is to the right so that the body moves to the right, past its equilibrium position, and hence reaches its initial displaced position. In practice this position will not quite be reached because damping in the system will have dissipated some of the vibrational energy. However, if the damping is small its effect can be neglected.

If the body is displaced a distance x_0 to the right and released, the free-body diagrams (FBDs) for a general displacement x are as shown in Fig. 2.2(a) and (b).

The effective force is always in the direction of positive x. If the body is being retarded $\ddot{x}$ will be calculated to be negative. The mass of the body is assumed constant: this is usually so but not always, as, for example, in the case of a rocket burning fuel. The spring stiffness k is assumed constant: this is usually so within limits (see section 2.1.3). It is assumed that the mass of the spring is negligible compared with the mass of the body; cases where this is not so are considered in section 2.1.4.1.

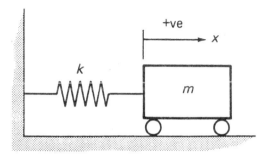

Fig. 2.1.   Single degree of freedom model – translation vibration.

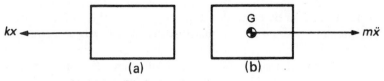

Fig. 2.2.   (a) Applied force; (b) effective force.

From the free-body diagrams the equation of motion for the system is

$$m\ddot{x} = - kx \quad \text{or} \quad \ddot{x} + (k/m)x = 0. \tag{2.1}$$

This will be recognized as the equation for simple harmonic motion. The solution is

$$x = A \cos \omega t + B \sin \omega t, \tag{2.2}$$

where A and B are constants which can be found by considering the initial conditions, and ω is the circular frequency of the motion. Substituting (2.2) into (2.1) we get

$$- \omega^2 (A \cos \omega t + B \sin \omega t) + (k/m) (A \cos \omega t + B \sin \omega t) = 0.$$

Since $(A \cos \omega t + B \sin \omega t) \neq 0$ (otherwise no motion),

$$\omega = \sqrt{(k/m)} \text{ rad/s,}$$

and

$$x = A \cos \sqrt{(k/m)}t + B \sin \sqrt{(k/m)}t.$$

Now

$$x = x_0 \text{ at } t = 0,$$

thus

$$x_0 = A \cos 0 + B \sin 0, \quad \text{and therefore } x_0 = A,$$

and

$$\dot{x} = 0 \text{ at } t = 0,$$

thus

$$0 = - A\sqrt{(k/m)} \sin 0 + B\sqrt{(k/m)} \cos 0, \quad \text{and therefore } B = 0;$$

that is,

$$x = x_0 \cos \sqrt{(k/m)}t. \tag{2.3}$$

The system parameters control ω and the type of motion but not the amplitude x_0, which is found from the initial conditions. The mass of the body is important, but its weight is not, so that for a given system, ω is independent of the local gravitational field.
    The frequency of vibration, f, is given by

$$f = \frac{\omega}{2\pi}, \quad \text{or} \quad f = \frac{1}{2\pi}\sqrt{\left(\frac{k}{m}\right)} \text{Hz.} \tag{2.4}$$

The motion is as shown in Fig. 2.3.

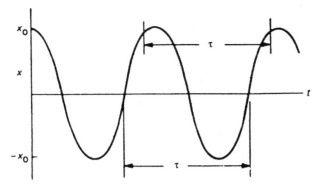

Fig. 2.3.   Simple harmonic motion.

The period of the oscillation, τ, is the time taken for one complete cycle so that

$$\tau = \frac{1}{f} = 2\pi\sqrt{(m/k)} \text{ seconds.} \tag{2.5}$$

The analysis of the vibration of a body supported to vibrate only in the vertical or y direction can be carried out in a similar way to that above.

It is found that for a given system the frequency of vibration is the same whether the body vibrates in a horizontal or vertical direction.

Sometimes more than one spring acts in a vibrating system. The spring, which is considered to be an elastic element of constant stiffness, can take many forms in practice; for example, it may be a wire coil, rubber block, beam or air bag. Combined spring units can be replaced in the analysis by a single spring of equivalent stiffness as follows.

2.1.1.1 Springs connected in series

The three-spring system of Fig. 2.4(a) can be replaced by the equivalent spring of Fig. 2.4(b).

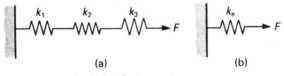

(a)   (b)

Fig. 2.4.   Spring systems.

If the deflection at the free end, δ, experienced by applying the force F is to be the same in both cases,

$$\delta = F/k_e = F/k_1 + F/k_2 + F/k_3,$$

that is,

$$1/k_e = \sum_1^3 1/k_i.$$

In general, the reciprocal of the equivalent stiffness of springs connected in series is obtained by summing the reciprocal of the stiffness of each spring.

2.1.1.2 Springs connected in parallel

The three-spring system of Fig. 2.5(a) can be replaced by the equivalent spring of Fig. 2.5(b).

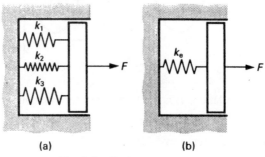

(a)   (b)

Fig. 2.5.   Spring systems.

Since the defection δ must be the same in both cases, the sum of the forces exerted by the springs in parallel must equal the force exerted by the equivalent spring. Thus

$$F = k_1\delta + k_2\delta + k_3\delta = k_e\delta,$$

that is,

$$k_e = \sum_{i=1}^{3} k_i.$$

In general, the equivalent stiffness of springs connected in parallel is obtained by summing the stiffness of each spring.

2.1.2 Torsional vibration

Fig. 2.6 shows the model used to study torsional vibration.

   A body with mass moment of inertia I about the axis of rotation is fastened to a bar of torsional stiffness k_T. If the body is rotated through an angle θ_0 and released, torsional vibration of the body results. The mass moment of inertia of the shaft about the axis of rotation is usually negligible compared with I.

   For a general displacement θ, the FBDs are as given in Fig. 2.7(a) and (b). Hence the equation of motion is

$$I\ddot{\theta} = -k_T\theta$$

or

$$\ddot{\theta} + \left(\frac{k_T}{I}\right)\theta = 0.$$

This is of a similar form to equation (2.1); that is, the motion is simple harmonic with frequency $(1/2\pi)\sqrt{(k_T/I)}$ Hz.

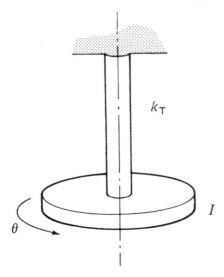

Fig. 2.6.   Single degree of freedom model – torsional vibration.

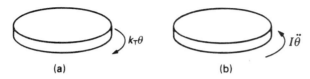

(a)   **(b)**

Fig. 2.7.   (a) Applied torque; (b) effective torque.

The torsional stiffness of the shaft, k_T, is equal to the applied torque divided by the angle of twist.

Hence

$$k_T = \frac{GJ}{l}, \text{ for a circular section shaft,}$$

where G = modulus of rigidity for shaft material,
   J = second moment of area about the axis of rotation, and
   l = length of shaft.

Hence

$$f = \frac{\omega}{2\pi} = \frac{1}{2\pi} \sqrt{(GJ/Il)} \text{ Hz,}$$

and

$$\theta = \theta_0 \cos \sqrt{(GJ/Il)}t,$$

when $\theta = \theta_0$ and $\dot{\theta} = 0$ at $t = 0$.

If the shaft does not have a constant diameter, it can be replaced analytically by an equivalent shaft of different length but with the same stiffness and a constant diameter.

For example, a circular section shaft comprising a length l_1 of diameter d_1 and a length l_2 of diameter d_2 can be replaced by a length l_1 of diameter d_1 and a length l of diameter d_1 where, for the same stiffness,

$$(GJ/l)_{\text{length } l_2 \text{ diameter } d_2} = (GJ/l)_{\text{length } l \text{ diameter } d_1}$$

that is, for the same shaft material, $d_2{}^4/l_2 = d_1{}^4/l$.

Therefore the equivalent length l_e of the shaft of constant diameter d_1 is given by

$$l_e = l_1 + (d_1/d_2)^4 l_2.$$

It should be noted that the analysis techniques for translational and torsional vibration are very similar, as are the equations of motion.

2.1.3 Non-linear spring elements

Any spring elements have a force–deflection relationship that is linear only over a limited range of deflection. Fig. 2.8 shows a typical characteristic.

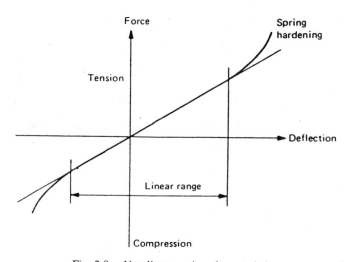

Fig. 2.8. Non-linear spring characteristic.

The non-linearities in this characteristic may be caused by physical effects such as the contacting of coils in a compressed coil spring, or by excessively straining the spring material so that yielding occurs. In some systems the spring elements do not act at the same time, as shown in Fig. 2.9 (a), or the spring is designed to be non-linear as shown in Fig. 2.9 (b) and (c).

Analysis of the motion of the system shown in Fig. 2.9 (a) requires analysing the motion until the half-clearance a is taken up, and then using the displacement and velocity at this point as initial conditions for the ensuing motion when the extra springs are operating. Similar analysis is necessary when the body leaves the influence of the extra springs.

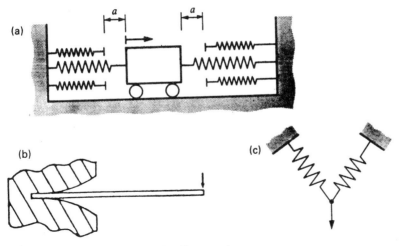

Fig. 2.9.   Non-linear spring systems.

2.1.4 Energy methods for analysis

For undamped free vibration the total energy in the vibrating system is constant throughout the cycle. Therefore the maximum potential energy V_{max} is equal to the maximum kinetic energy T_{max} although these maxima occur at different times during the cycle of vibration. Furthermore, since the total energy is constant,

$$T + V = \text{constant},$$

and thus

$$\frac{d}{dt}(T + V) = 0.$$

Applying this method to the case, already considered, of a body of mass m fastened to a spring of stiffness k, when the body is displaced a distance x from its equilibrium position,

strain energy (SE) in spring $= \frac{1}{2}kx^2$.
kinetic energy (KE) of body $= \frac{1}{2}m\dot{x}^2$.

Hence

$$V = \tfrac{1}{2}kx^2,$$

and

$$T = \tfrac{1}{2}m\dot{x}^2.$$

Thus

$$\frac{d}{dt}(\tfrac{1}{2}m\dot{x}^2 + \tfrac{1}{2}kx^2) = 0,$$

that is

$$m\ddot{x} + k\dot{x}x = 0,$$

or

$$\ddot{x} + \left(\frac{k}{m}\right)x = 0, \text{ as before in equation (2.1).}$$

This is a very useful method for certain types of problem in which it is difficult to apply Newton's laws of motion.

Alternatively, assuming SHM, if $x = x_0 \cos \omega t$,

the maximum SE, $V_{max} = \frac{1}{2}kx_0^2$,

and

the maximum KE, $T_{max} = \frac{1}{2}m(x_0\omega)^2$.

Thus, since $T_{max} = V_{max}$,

$$\tfrac{1}{2}kx_0^2 = \tfrac{1}{2}mx_0^2\omega^2,$$

or $\omega = \sqrt{(k/m)}$ rad/s.

Energy methods can also be used in the analysis of the vibration of continuous systems such as beams. It has been shown by Rayleigh that the lowest natural frequency of such systems can be fairly accurately found by assuming any reasonable deflection curve for the vibrating shape of the beam: this is necessary for the evaluation of the kinetic and potential energies. In this way the continuous system is modelled as a single degree of freedom system, because once one coordinate of beam vibration is known, the complete beam shape during vibration is revealed. Naturally the assumed deflection curve must be compatible with the end conditions of the system, and since any deviation from the true mode shape puts additional constraints on the system, the frequency determined by Rayleigh's method is never less than the exact frequency. Generally, however, the difference is only a few per cent. The frequency of vibration is found by considering the conservation of energy in the system; the natural frequency is determined by dividing the expression for potential energy in the system by the expression for kinetic energy.

2.1.4.1 The vibration of systems with heavy springs

The mass of the spring element can have a considerable effect on the frequency of vibration of those structures in which heavy springs are used.

Consider the translational system shown in Fig. 2.10, where a rigid body of mass M is connected to a fixed frame by a spring of mass m, length l, and stiffness k. The body moves in the x direction only. If the dynamic deflected shape of the spring is assumed to be the same as the static shape, the velocity of the spring element is $\dot{y} = (y/l)\dot{x}$, and the mass of the element is $(m/l)dy$.

Thus

$$T = \tfrac{1}{2}M\dot{x}^2 + \int_0^l \frac{1}{2}\left(\frac{m}{l}\right)\left[\frac{y}{l}\dot{x}\right]^2 dy$$

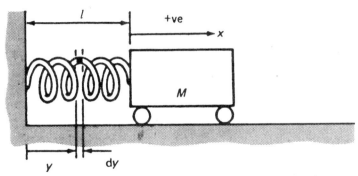

Fig. 2.10.   Single degree of freedom system with heavy spring.

$$= \frac{1}{2}\left(M + \frac{m}{3}\right)\dot{x}^2$$

and

$$V = \tfrac{1}{2}kx^2.$$

Assuming simple harmonic motion and putting $T_{\max} = V_{\max}$ gives the frequency of free vibration as

$$f = \frac{1}{2\pi}\sqrt{\left(\frac{k}{M + (m/3)}\right)}\ \text{Hz},$$

that is, if the system is to be modelled with a massless spring, one third of the actual spring mass must be added to the mass of the body in the frequency calculation.

Alternatively, $\dfrac{d}{dt}(T + V) = 0$ can be used for finding the frequency of oscillation.

2.1.4.2 Transverse vibration of beams

For the beam shown in Fig. 2.11, if m is the mass unit length and y is the amplitude of the assumed deflection curve, then

$$T_{\max} = \tfrac{1}{2}\int \dot{y}^2_{\max}\, dm = \tfrac{1}{2}\omega^2 \int y^2\, dm,$$

where ω is the natural circular frequency of the beam.

The strain energy of the beam is the work done on the beam which is stored as elastic energy. If the bending moment is M and the slope of the elastic curve is θ,

$$V = \tfrac{1}{2}\int M\, d\theta.$$

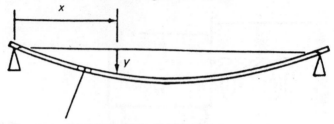

Beam segment shown enlarged below

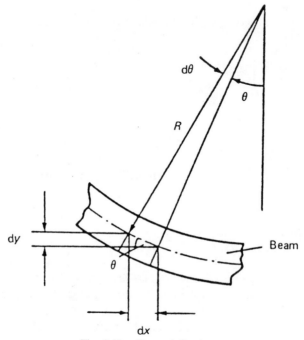

Fig. 2.11.   Beam deflection.

Usually the deflection of beams is small so that the following relationships can be assumed to hold:

$$\theta = \frac{dy}{dx} \quad \text{and} \quad R\,d\theta = dx;$$

thus

$$\frac{1}{R} = \frac{d\theta}{dx} = \frac{d^2y}{dx^2}.$$

From beam theory, $M/I = E/R$, where R is the radius of curvature and EI is the flexural rigidity. Thus

$$V = \tfrac{1}{2} \int \frac{M}{R} dx = \tfrac{1}{2} \int EI \left(\frac{d^2 y}{dx^2}\right)^2 dx.$$

Since

$$T_{max} = V_{max};$$

$$\omega^2 = \frac{\displaystyle\int EI \left(\frac{d^2 y}{dx^2}\right)^2 dx}{\displaystyle\int y^2 \, dm}.$$

This expression gives the lowest natural frequency of transverse vibration of a beam. It can be seen that to analyse the transverse vibration of a particular beam by this method requires y to be known as a function of x. For this the static deflected shape or a part sinusoid can be assumed, provided the shape is compatible with the beam boundary conditions.

2.1.5 The stability of vibrating structures

If a structure is to vibrate about an equilibrium position, it must be stable about that position; that is, if the structure is disturbed when in an equilibrium position, the elastic forces must be such that the structure vibrates about the equilibrium position. Thus the expression for ω^2 must be positive if a real value of the frequency of vibration about the equilibrium position is to exist, and hence the potential energy of a stable structure must also be positive.

The principle of minimum potential energy can be used to test the stability of structures that are conservative. Thus a structure will be stable at an equilibrium position if the potential energy of the structure is a minimum at that position. This requires that

$$\frac{dV}{dq} = 0 \quad \text{and} \quad \frac{d^2 V}{dq^2} > 0$$

where q is an independent or generalized coordinate. Hence the necessary conditions for vibration to take place are found, and the position about which the vibration occurs is determined.

Example 1

A link AB in a mechanism is a rigid bar of uniform section 0.3 m long. It has a mass of 10 kg, and a concentrated mass of 7 kg is attached at B. The link is hinged at A and is supported in a horizontal position by a spring attached at the mid-point of the bar. The stiffness of the spring is 2 kN/m. Find the frequency of small free oscillations of the system. The system is as follows.

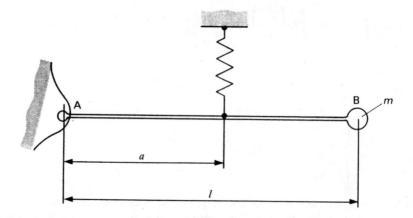

For rotation about A the equation of motion is

$$I_A\ddot{\theta} = -ka^2\theta$$

that is,

$$\ddot{\theta} + (ka^2/I_A)\theta = 0.$$

This is SHM with frequency

$$\frac{1}{2\pi}\sqrt{(ka^2/I_A)} \text{ Hz.}$$

In this case

$$a = 0.15 \text{ m}, \, l = 0.3 \text{ m}, \, k = 2000 \text{ N/m},$$

and

$$I_A = 7(0.3)^2 + \tfrac{1}{3} \times 10 \, (0.3)^2 = 0.93 \text{ kg m}^2.$$

Hence

$$f = \frac{1}{2\pi}\sqrt{\left(\frac{2000(0.15)^2}{0.93}\right)} = 1.1 \text{ Hz.}$$

Example 2

A uniform cylinder of mass m is rotated through a small angle θ_0 from the equilibrium position and released. Determine the equation of motion and hence obtain the frequency of free vibration. The cylinder rolls without slipping.

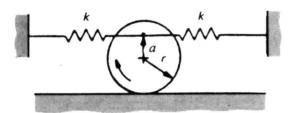

If the axis of the cylinder moves a distance x and turns through an angle θ so that $x = r\theta$

$$\text{KE} = \tfrac{1}{2}m\dot{x}^2 + \tfrac{1}{2}I\dot{\theta}^2, \text{ where } I = \tfrac{1}{2}mr^2.$$

Hence

$$\text{KE} = \tfrac{3}{4}mr^2\dot{\theta}^2.$$
$$\text{SE} = 2 \times \tfrac{1}{2} \times k[(r + a)\theta]^2 = k(r + a)^2\theta^2.$$

Now, energy is conserved, so $(\tfrac{3}{4}mr^2\dot{\theta}^2 + k(r + a)^2\theta^2)$ is constant; that is,

$$\frac{d}{dt}(\tfrac{3}{4}mr^2\dot{\theta}^2 + k(r + a)^2\theta^2) = 0$$

or

$$\tfrac{3}{4}mr^2 2\dot{\theta}\ddot{\theta} + k(r + a)^2 2\theta\dot{\theta} = 0.$$

Thus the equation of the motion is

$$\ddot{\theta} + \frac{k(r + a)^2\theta}{(\tfrac{3}{4})mr^2} = 0.$$

Hence the frequency of free vibration is

$$\frac{1}{2\pi}\sqrt{\left[\frac{4k(r + a)^2}{3mr^2}\right]} \text{ Hz.}$$

Example 3

A uniform wheel of radius R can roll without slipping on an inclined plane. Concentric with the wheel, and fixed to it, is a drum of radius r around which is wrapped one end of a string. The other end of the string is fastened to an anchored spring, of stiffness k, as shown. Both spring and string are parallel to the plane. The total mass of the wheel/drum assembly is m and its moment of inertia about the axis through the centre of the wheel O is I. If the wheel is displaced a small distance from its equilibrium position and released, derive the equation describing the ensuing motion and hence calculate the frequency of the oscillations. Damping is negligible.

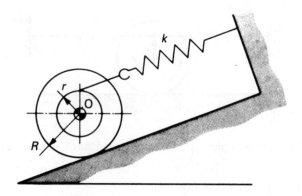

If the wheel is given an anti-clockwise rotation θ from the equilibrium position, the spring extension is $(R + r)\,\theta$ so that the restoring spring force is $k(R + r)\theta$.

The FBDs are:

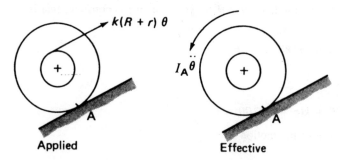

$k(R + r)\,\theta$

$I_A\ddot{\theta}$

Applied   Effective

The rotation is instantaneously about the contact point A so that taking moments about A gives the equation of motion as

$$I_A\ddot{\theta} = -k(R + r)^2\theta.$$

(The moment due to the weight cancels with the moment due to the initial spring tension.)

Now $I_A = I + mR^2$, so

$$\ddot{\theta} + \left(\frac{k(R + r)^2}{I + mR^2}\right)\theta = 0,$$

and the frequency of oscillation is

$$\frac{1}{2\pi}\sqrt{\left(\frac{k(R + r)^2}{I + mR^2}\right)}\ \text{Hz.}$$

An alternative method for obtaining the frequency of oscillation is to consider the energy in the system.

Now

SE, $V = \frac{1}{2}k(R + r)^2\theta^2$,

and

KE, $T = \frac{1}{2}I_A\dot{\theta}^2$,

(weight and initial spring tension effects cancel) so

$$T + V = \frac{1}{2}I_A\dot{\theta}^2 + \frac{1}{2}k(R + r)^2\theta^2,$$

and

$$\frac{d}{dt}(T + V) = \frac{1}{2}I_A2\dot{\theta}\ddot{\theta} + \frac{1}{2}k(R + r)^2 2\theta\dot{\theta} = 0.$$

Hence

$$I_A\ddot{\theta} + k(R + r)^2\theta = 0,$$

which is the equation of motion.

Or, we can put $V_{max} = T_{max}$, and if $\theta = \theta_0 \sin \omega t$ is assumed,

$$\frac{1}{2}k(R + r)^2\theta_0^2 = \frac{1}{2}I_A\omega^2\theta_0^2,$$

so that

$$\omega = \sqrt{\left(\frac{k(R + r)^2}{I_A}\right)} \text{ rad/s,}$$

where

$$I_A = I + mr^2 \quad \text{and} \quad f = (\omega/2\pi) \text{ Hz.}$$

Example 4

A simply supported beam of length l and mass m_2 carries a body of mass m_1 at its mid-point. Find the lowest natural frequency of transverse vibration.

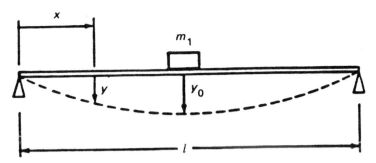

The boundary conditions are $y = 0$ and $d^2y/dx^2 = 0$ at $x = 0$ and $x = l$. These conditions are satisfied by assuming that the shape of the vibrating beam can be represented by a half sine wave. A polynomial expression can be derived for the deflected shape, but the sinusoid is usually easier to manipulate.

$y = y_0 \sin(\pi x/l)$ is a convenient expression for the beam shape, which agrees with the boundary conditions. Now

$$\dot{y} = \dot{y}_0 \sin\left(\frac{\pi x}{l}\right) \quad \text{and} \quad \frac{d^2 y}{dx^2} = -y_0 \left(\frac{\pi}{l}\right)^2 \sin\left(\frac{\pi x}{l}\right).$$

Hence

$$\int_0^l EI\left(\frac{d^2 y}{dx^2}\right)^2 dx = \int_0^l EI y_0^2 \left(\frac{\pi}{l}\right)^4 \sin^2\left(\frac{\pi x}{l}\right) dx$$

$$= EI y_0^2 \left(\frac{\pi}{l}\right)^4 \frac{l}{2},$$

and

$$\int y^2 \, dm = \int_0^l y_0^2 \sin^2\left(\frac{\pi x}{l}\right) \frac{m_2}{l} dx + y_0^2 m_1$$

$$= y_0^2 \left(m_1 + \frac{m_2}{2}\right). \quad \text{Thus}$$

$$\omega^2 = \frac{EI(\pi/l)^4 \, l/2}{(m_1 + m_2/2)}.$$

If $m_2 = 0$,

$$\omega^2 = \frac{EI}{2} \frac{\pi^4}{l^3 m_1} = 48.7 \frac{EI}{m_1 l^3}.$$

The exact solution is $48 \, EI/m_1 l^3$, so the Rayleigh method solution is 1.4% high.

Example 5

Find the lowest natural frequency of transverse vibration of a cantilever of mass m, which has rigid body of mass M attached at its free end.

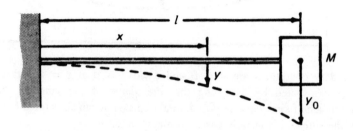

The static deflection curve is $y = (y_0/2l^3)(3lx^2 - x^3)$. Alternatively $y = y_0(1 - \cos \pi x/2l)$ could be assumed. Hence

$$\int_0^l EI\left(\frac{d^2y}{dx^2}\right)^2 dx = EI\int_0^l \left(\frac{y_0}{2l^3}\right)^2 (6l - 6x)^2 \, dx = \frac{3EI}{l^3}y_0^2,$$

and

$$\int y^2 \, dm = \int_0^l y^2 \frac{m}{l} \, dx + y_0^2 M$$

$$= \int_0^l \frac{y_0^2}{4l^6} \frac{m}{l} (3lx^2 - x^3)^2 \, dx + y_0^2 M$$

$$= y_0^2\left\{M + \frac{33}{140}m\right\}.$$

Thus

$$\omega^2 = \left\{\frac{3EI}{\left(M + \dfrac{33}{140}m\right)l^3}\right\} \ (\text{rad/s})^2.$$

Example 6

Part of an industrial plant incorporates a horizontal length of uniform pipe, which is rigidly embedded at one end and is effectively free at the other. Considering the pipe as a cantilever, derive an expression for the frequency of the first mode of transverse vibration using Rayleigh's method.

   Calculate this frequency, given the following data for the pipe:

Modulus of elasticity	$200 \ \text{GN/m}^2$
Second moment of area about bending axis	$0.02 \ \text{m}^4$
Mass	$6 \times 10^4 \ \text{kg}$
Length	30 m
Outside diameter	1 m

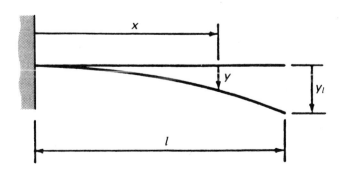

For a cantilever, assume

$$y = y_I\left(1 - \cos\frac{\pi x}{2l}\right).$$

This is compatible with zero deflection and slope when $x = 0$, and zero shear force and bending moment when $x = l$. Thus

$$\frac{d^2y}{dx^2} = y_I\left(\frac{\pi}{2l}\right)^2 \cos\frac{\pi x}{2l}.$$

Now

$$\int_0^l EI\left(\frac{d^2y}{dx^2}\right)^2 dx = EI\int_0^l y_I^2\left(\frac{\pi}{2l}\right)^4 \cos^2\frac{\pi x}{2l}\,dx$$

$$= EIy_I^2\left(\frac{\pi}{2l}\right)^4 \frac{l}{2}$$

and

$$\int_0^l y^2\,dm = \int_0^l y_I^2\left(1 - \cos\frac{\pi x}{2l}\right)^2 \frac{m}{l}\,dx$$

$$= y_I^2 m\left(\frac{3}{2} - \frac{4}{\pi}\right).$$

Hence, assuming the structure to be conservative, that is, the total energy remains constant throughout the vibration cycle,

$$\omega^2 = \frac{EIy_I^2\left(\dfrac{\pi}{2l}\right)^4 \dfrac{l}{2}}{y_I^2 m\left(\dfrac{3}{2} - \dfrac{4}{\pi}\right)}$$

$$= \frac{EI}{ml^3}\,13.4.$$

Thus

$$\omega = 3.66\sqrt{\left(\frac{EI}{ml^3}\right)}\text{ rad/s}\quad\text{and}\quad f = \frac{3.66}{2\pi}\sqrt{\left(\frac{EI}{ml^3}\right)}\text{ Hz.}$$

In this case

$$\frac{EI}{ml^3} = \frac{200 \times 10^9 \times 0.02}{6 \times 10^4 \times 30^3} |s^2.$$

Hence

$$\omega = 5.75 \text{ rad/s} \quad \text{and} \quad f = 0.92 \text{ Hz}.$$

Example 7

A uniform building of height $2h$ and mass m has a rectangular base $a \times b$ which rests on an elasic soil. The stiffness of the soil, k, is expressed as the force per unit area required to produce unit deflection.
Find the lowest frequency of free low-amplitude swaying oscillation of the building.

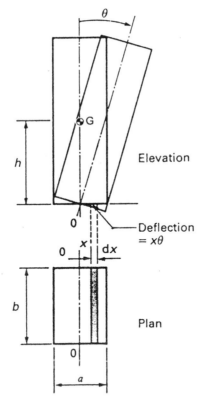

The lowest frequency of oscillation about the axis O–O through the base of the building is when the oscillation occurs about the shortest side, of length a.
I_o is the mass moment of inertia of the building about axis O–O.

The FBDs are:

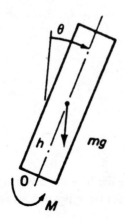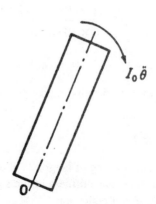

and the equation of motion for small θ is given by

$$I_0\ddot{\theta} = mgh\theta - M,$$

where M is the restoring moment from the elastic soil.

For the soil, $k = $ force/(area × deflection), so considering an element of the base as shown, the force on element $= kb\,dx \times x\theta$, and the moment of this force about axis O–O $= kb\,dx \times x\theta x$. Thus the total restoring moment M, assuming the soil acts similarly in tension and compression, is

$$M = 2\int_0^{a/2} kbx^2\theta\,dx$$

$$= 2kb\theta\frac{(a/2)^3}{3} = \frac{ka^3b}{12}\theta.$$

Thus the equation of motion becomes

$$I_0\ddot{\theta} + \left(\frac{ka^3b}{12} - mgh\right)\theta = 0.$$

Motion is therefore simple harmonic, with frequency

$$f = \frac{1}{2\pi}\sqrt{\left(\frac{ka^3b/12 - mgh}{I_0}\right)}\ \text{Hz.}$$

An alternative solution can be obtained by considering the energy in the system. In this case,

$$T = \tfrac{1}{2}I_0\dot{\theta}^2,$$

and

$$V = \tfrac{1}{2}.2\int_0^{a/2} kb\,dx \times x\theta \times x\theta - \frac{mgh\theta^2}{2},$$

where the loss in potential energy of the building weight is given by $mgh (1 - \cos \theta) \simeq mgh\theta^2/2$, since $\cos \theta \simeq 1 - \theta^2/2$ for small values of θ. Thus

$$V = \left(\frac{ka^3b}{24} - \frac{mgh}{2} \right) \theta^2.$$

Assuming simple harmonic motion, and putting $T_{max} = V_{max}$, gives

$$\omega^2 = \left(\frac{ka^3b/12 - mgh}{I_0} \right)$$

as before.

Note that for stable oscillation, $\omega > 0$, so that

$$\left(\frac{ka^3b}{12} - mgh \right) > 0,$$

that is, $ka^3b > 12mgh$.

This expression gives the minimum value of k, the soil stiffness, for stable oscillation of a particular building to occur. If k is less that $12\, mgh/a^3b$ the building will fall over when disturbed.

2.2 FREE DAMPED VIBRATION

All real structures dissipate energy when they vibrate. The energy dissipated is often very small, so that an undamped analysis is sometimes realistic; but when the damping is significant its effect must be included in the analysis, particularly when the amplitude of vibration is required. Energy is dissipated by frictional effects, for example that occurring at the connection between elements, internal friction in deformed members, and windage. It is often difficult to model damping exactly because many mechanisms may be operating in a structure. However, each type of damping can be analysed, and since in many dynamic systems one form of damping predominates, a reasonably accurate analysis is usually possible.

The most common types of damping are viscous, dry friction and hysteretic. Hysteretic damping arises in structural elements due to hysteresis losses in the material.

The type and amount of damping in a structure has a large effect on the dynamic response levels.

2.2.1 Vibration with viscous damping

Viscous damping is a common form of damping which is found in many engineering systems such as instruments and shock absorbers. The viscous damping force is proportional to the first power of the velocity across the damper, and it always opposes the motion, so that the damping force is a linear continuous function of the velocity. Because the analysis of viscous damping leads to the simplest mathematical treatment, analysts sometimes approximate more complex types of damping to the viscous type.

Consider the single degree of freedom model with viscous damping shown in Fig. 2.12.

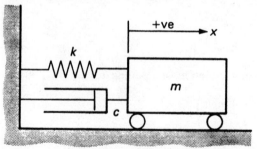

Fig. 2.12.   Single degree of freedom model with viscous damping.

The only unfamiliar element in the system is the viscous damper with coefficient c. This coefficient is such that the damping force required to move the body with a velocity $\dot{x}$ is $c\dot{x}$.

For motion of the body in the direction shown, the free body diagrams are as in Fig. 2.13(a) and (b).

Fig. 2.13.   (a) Applied force; (b) effective force.

The equation of motion is therefore

$$m\ddot{x} + c\dot{x} + kx = 0. \tag{2.6}$$

This equation of motion pertains to the whole of the cycle: the reader should verify that this is so. (Note: displacements to the left of the equilibrium position are negative, and velocities and accelerations from right to left are also negative.)

Equation (2.6) is a second-order differential equation which can be solved by assuming a solution of the form $x = Xe^{st}$. Substituting this solution into equation (2.6) gives

$$(ms^2 + cs + k)Xe^{st} = 0.$$

Since $Xe^{st} \neq 0$ (otherwise no motion),

$$ms^2 + cs + k = 0.$$

If the roots of the equation are s_1 and s_2, then

$$s_{1,2} = -\frac{c}{2m} \pm \frac{\sqrt{(c^2 - 4mk)}}{2m}.$$

Hence

$$x = X_1 e^{s_1 t} + X_2 e^{s_2 t},$$

where X_1 and X_2 are arbitrary constants found from the initial conditions. The system response evidently depends upon whether c is positive or negative, and on whether c^2 is greater than, equal to, or less than $4mk$.

The dynamic behaviour of the system depends upon the numerical value of the radical, so we define critical damping as that value of $c(c_c)$ which makes the radical zero; that is,

$$c_c = 2\sqrt{(km)}.$$

Hence

$$c_c/2m = \sqrt{(k/m)} = \omega, \quad \text{the undamped natural frequency,}$$

and

$$c_c = 2\sqrt{(km)} = 2m\omega.$$

The actual damping in a system can be specified in terms of c_c by introducing the damping ratio ζ.

Thus

$$\zeta = c/c_c,$$

and

$$s_{1,2} = [-\zeta \pm \sqrt{(\zeta^2 - 1)}]\omega. \tag{2.7}$$

The response evidently depends upon whether c is positive or negative, and upon whether ζ is greater than, equal to, or less than unity. Usually c is positive, so we only need to consider the other possibilities.

Case 1. $\zeta < 1$; that is, damping less than critical

From equation (2.7)

$$s_{1,2} = -\zeta\omega \pm j\sqrt{(1 - \zeta^2)}\omega, \quad \text{where } j = \sqrt{(-1)},$$

so

$$x = e^{-\zeta\omega t}[X_1 e^{j\sqrt{(1-\zeta^2)}\omega t} + X_2 e^{-j\sqrt{(1-\zeta^2)}\omega t}]$$

and

$$x = Xe^{-\zeta\omega t} \sin (\sqrt{(1 - \zeta^2)}\,\omega t + \phi).$$

The motion of the body is therefore an exponentially decaying harmonic oscillation with circular frequency $\omega_v = \omega\sqrt{(1 - \zeta^2)}$, as shown in Fig. 2.14.

The frequency of the viscously damped oscillation ω_v, is given by $\omega_v = \omega\sqrt{(1 - \zeta^2)}$, that is, the frequency of oscillation is reduced by the damping action. However, in many systems this reduction is likely to be small, because very small values of ζ are common; for example, in most engineering structures ζ is rarely greater than 0.02. Even if $\zeta = 0.2$, $\omega_v = 0.98\omega$. This is not true in those cases where ζ is large, for example in motor vehicles where ζ is typically 0.7 for new shock absorbers.

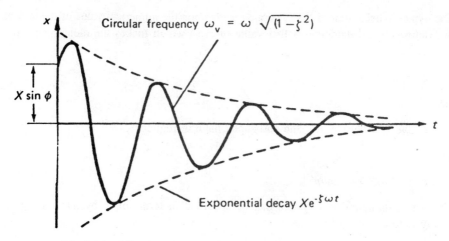

Fig. 2.14. Vibration decay of system with viscous damping, $\zeta < 1$.

Case 2. $\zeta = 1$; that is, critical damping

Both values of s are $-\omega$. However, two constants are required in the solution of equation (2.6); thus $x = (A + Bt)e^{-\omega t}$ may be assumed.

Critical damping represents the limit of periodic motion; hence the displaced body is restored to equilibrium in the shortest possible time, and without oscillation or overshoot. Many devices, particularly electrical instruments, are critically damped to take advantage of this property.

Case 3. $\zeta > 1$; that is, damping greater than critical

There are two real values of s, so $x = X_1 e^{s_1 t} + X_2 e^{s_2 t}$.

Since both values of s are negative the motion is the sum of two exponential decays, as shown in Fig. 2.15.

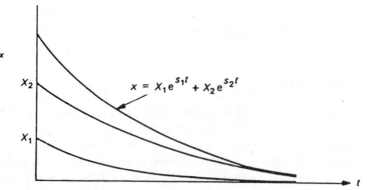

Fig. 2.15. Disturbance decay of system with viscous damping $\zeta > 1$.

2.2.1.1 Logarithmic decrement Λ

A convenient way of determining the damping in a system is to measure the rate of decay
of oscillation. It is usually not satisfactory to measure ω_v and ω because unless $\zeta > 0.2$,
$\omega \simeq \omega_v$.

The logarithmic decrement, Λ, is the natural logarithm of the ratio of any two
successive amplitudes in the same direction, and so from Fig. 2.16

$$\Lambda = \ln\frac{X_1}{X_{11}}$$

where X_1 and X_{11} are successive amplitudes as shown.
Since

$$x = Xe^{-\zeta\omega t} \sin(\omega_v t + \phi),$$

if

$$X_1 = Xe^{-\zeta\omega t}, \quad \text{then} \quad X_{11} = Xe^{-\zeta\omega(t + \tau_v)},$$

where τ_v is the period of the damped oscillation.

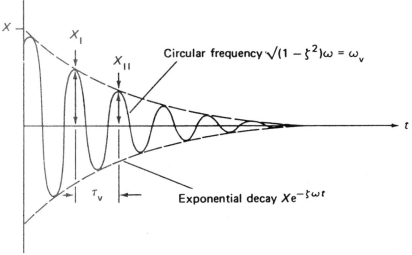

Fig. 2.16.   Vibration decay.

Thus

$$\Lambda = \ln\frac{Xe^{-\zeta\omega t}}{Xe^{-\zeta\omega(t + \tau_v)}} = \zeta\omega\tau_v.$$

Since

$$\tau_v = \frac{2\pi}{\omega_v} = \frac{2\pi}{\omega\sqrt{(1 - \zeta^2)}},$$

then

$$\Lambda = \frac{2\pi\zeta}{\sqrt{(1 - \zeta^2)}} = \ln\left(\frac{X_1}{X_{11}}\right).$$

For small values of $\zeta(\not> 0.25)$, $\Lambda \simeq 2\pi\zeta$.

It should be noted that this analysis assumes that the point of maximum displacement in a cycle and the point where the envelope of the decay curve $Xe^{-\zeta\omega t}$ touches the decay curve itself, are coincident. This is usually very nearly so, and the error in making this assumption is usually negligible, except in those cases where the damping is high.

For low damping it is preferable to measure the amplitude of oscillations many cycles apart so that an easily measurable difference exists.

In this case

$$\Lambda = \ln\left(\frac{X_1}{X_{11}}\right) = \frac{1}{N}\ln\left(\frac{X_1}{X_{N+1}}\right)$$

since

$$\frac{X_1}{X_{11}} = \frac{X_{11}}{X_{111}}, \text{ etc.}$$

Example 8

Consider the transverse vibration of a bridge structure. For the fundamental frequency it can be considered as a single degree of freedom system. The bridge is deflected at mid-span (by winching the bridge down) and suddenly released. After the initial disturbance the vibration was found to decay exponentially from an amplitude of 10 mm to 5.8 mm in three cycles with a frequency of 1.62 Hz. The test was repeated with a vehicle of mass 40 000 kg at mid-span, and the frequency of free vibration was measured to be 1.54 Hz.

Find the effective mass, the effective stiffness, and the damping ratio of the structure.

Let m be the effective mass and k the effective stiffness. Then

$$f_1 = 1.62 = \frac{1}{2\pi}\sqrt{\left(\frac{k}{m}\right)} \text{ Hz,}$$

and

$$f_2 = 1.54 = \frac{1}{2\pi}\sqrt{\left(\frac{k}{m + 40 \times 10^3}\right)} \text{ Hz,}$$

if it is assumed that ζ is small enough for $f_v \simeq f$.

Thus $\left(\frac{1.62}{1.54}\right)^2 = \frac{m + 40 \times 10^3}{m},$

so

$$m = 375 \times 10^3 \text{ kg.}$$

Since

$$k = (2\pi f_1)^2 m,$$
$$k = 38\ 850 \text{ kN/m.}$$

Now

$$\Lambda = \ln \frac{X_I}{X_{II}} = \tfrac{1}{3} \ln \frac{X_I}{X_{IV}} = \tfrac{1}{3} \ln \left(\frac{10}{5.8} \right)$$

$$= 0.182.$$

where X_I, X_{II} and X_{IV} are the amplitudes of the first, second and fourth cycles, respectively. Hence

$$\Lambda = \frac{2\pi\zeta}{\sqrt{(1 - \zeta^2)}} = 0.182,$$

and so $\zeta = 0.029$. (This compares with a value of about 0.002 for cast iron material. The additional damping originates mainly in the joints of the structure.) This value of ζ confirms the assumption that $f_v \simeq f$.

Example 9

A light rigid rod of length L is pinned at one end O and has a body of mass m attached at the other end. A spring and viscous damper connected in parallel are fastened to the rod at a distance a from the support. The system is set up in a *horizontal* plane: a plan view is shown.

Assuming that the damper is adjusted to provide critical damping, obtain the motion of the rod as a function of time if it is rotated through a small angle θ_0 and then released. Given that $\theta_0 = 2°$ and the undamped natural frequency of the system is 3 rad/s, calculate the displacement 1 s after release.

Explain the term *logarithmic decrement* as applied to such a system and calculate its value assuming that the damping is reduced to 80% of its critical value.

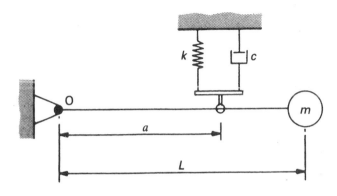

Let the rod turn through an angle θ from the equilibrium position. Note that the system oscillates in the horizontal plane so that the FBDs are:

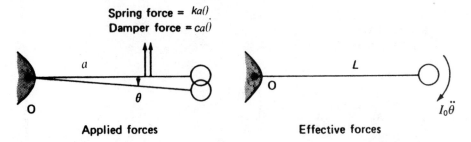

Spring force = $ka()$
Damper force = $ca()$

Applied forces   Effective forces

Taking moments about the pivot O gives

$$I_o\ddot{\theta} = -ca^2\dot{\theta} - ka^2\theta,$$

where $I_0 = mL^2$, so the equation of motion is

$$mL^2\ddot{\theta} + ca^2\dot{\theta} + ka^2\theta = 0.$$

Now the system is adjusted for critical damping, so that $\zeta = 1$. The solution to the equation of motion is therefore of the form

$$\theta = (A + Bt)e^{-\omega t}.$$

Now, $\theta = \theta_0$ when $t = 0$, and $d\theta/dt = 0$ when $t = 0$. Hence

$$\theta_0 = A,$$

and

$$0 = Be^{-\omega t} + (A + Bt)(-\omega)e^{-\omega t},$$

so that

$$B = \theta_0\omega.$$

Hence

$$\theta = \theta_0(1 + \omega)e^{-\omega t}.$$

If $\omega = 3$ rad/s, $t = 1$ s and $\theta_0 = 2°$,

$$\theta = 2(1 + 3)e^{-3} = 0.4°.$$

The logarithmic decrement

$$\Lambda = \ln\frac{X_1}{X_{11}} = \frac{2\pi\zeta}{\sqrt{(1 - \zeta^2)}},$$

so that if $\zeta = 0.8$,

$$\Lambda = \frac{5.027}{0.6} = 8.38$$

2.2.2 Vibration with Coulomb (dry friction) damping

Steady friction forces occur in many structures when relative motion takes place between adjacent members. These forces are independent of amplitude and frequency; they always oppose the motion and their magnitude may, to a first approximation, be considered constant. Dry friction can, of course, just be one of the damping mechanisms present; however, in some structures it is the main source of damping. In these cases the damping can be modelled as in Fig. 2.17.

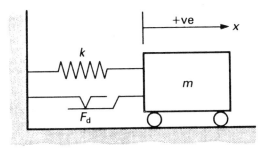

Fig. 2.17.   System with Coulomb damping.

The constant friction force F_d always opposes the motion, so that if the body is displaced a distance x_0 to the right and released from rest we have, for motion from right to left only,

$$m\ddot{x} = F_d - kx$$

or

$$m\ddot{x} + kx = F_d. \tag{2.8}$$

The solution to the complementary function is $x = A \sin \omega t + B \cos \omega t$, and the complete solution is

$$x = A \sin \omega t + B \cos \omega t + \frac{F_d}{k} \tag{2.9}$$

where $\omega = \sqrt{(k/m)}$ rad/s.

Note. The particular integral may be found by using the D-operator. Thus equation (2.8) is

$$(D^2 + \omega^2)x = F_d/m$$

so

$$x = (1/\omega^2)[1 + (D^2/\omega^2)]^{-1}F_d/m$$
$$= [1 - (D^2/\omega^2) + \cdots]F_d/m\omega^2 = F_d/k.$$

The initial conditions were $x = x_0$ at $t = 0$, and $\dot{x} = 0$ at $t = 0$. Substitution into equation (2.9) gives

$$A = 0 \quad \text{and} \quad B = x_0 - \frac{F_d}{k}.$$

Hence

$$x = \left(x_0 - \frac{F_d}{k}\right)\cos \omega t + \frac{F_d}{k}. \tag{2.10}$$

At the end of the half cycle right to left, $\omega t = \pi$ and

$$x_{(t = \pi/\omega)} = -x_0 + \frac{2F_d}{k},$$

that is, there is a reduction in amplitude of $2F_d/k$ per half cycle.

From symmetry, for motion from left to right when the friction force acts in the opposite direction to the above, the initial displacement is $(x_0 - 2F_d/k)$ and the final displacement is therefore $(x_0 - 4F_d/k)$, that is, the reduction in amplitude is $4F_d/k$ per cycle. This oscillation continues until the amplitude of the motion is so small that the maximum spring force is unable to overcome the friction force F_d. This can happen whenever the amplitude is $\leqslant \pm(F_d/k)$. The manner of oscillation decay is shown in Fig. 2.18; the motion is sinusoidal for each half cycle, with successive half cycles centred on points distant $+ (F_d/k)$ and $- (F_d/k)$ from the origin. The oscillation ceases with $|x| \leqslant F_d/k$. The zone $x = \pm F_d/k$ is called the *dead zone*.

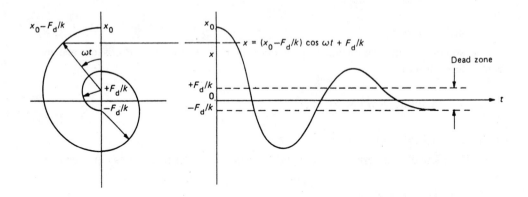

Fig. 2.18.   Vibration decay of system with Coulomb damping.

To determine the frequency of oscillation we rewrite the equation of motion (2.8) as

$$m\ddot{x} + k(x - (F_d/k)) = 0.$$

Now if $x' = x - (F_d/k)$, $\ddot{x}' = \ddot{x}$ so that $m\ddot{x}' + kx' = 0$, from which the frequency of oscillation is $(1/2\pi)\sqrt{(k/m)}$ Hz; that is, the frequency of oscillation is not affected by Coulomb friction.

Example 10

Part of a structure can be modelled as a torsional system comprising a bar of stiffness 10 kN m/rad and a beam of moment of inertia about the axis of rotation of 50 kg m². The bottom guide imposes a friction torque of 10 N m (see figure).

Given that the beam is displaced through 0.05 rad from its equilibrium position and released, find the frequency of the oscillation, the number of cycles executed before the beam motion ceases, and the position of the beam when this happens.

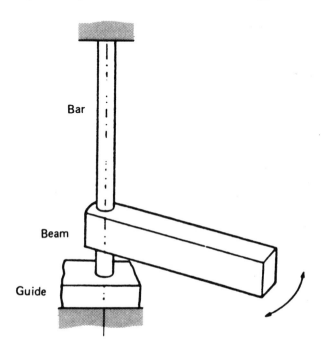

Now

$$\omega = \sqrt{\left(\frac{k_T}{I}\right)} = \sqrt{\left(\frac{10 \times 10^3}{50}\right)} = 14.14 \text{ rad/s.}$$

Hence

$$f = \frac{14.14}{2\pi} = 2.25 \text{ Hz.}$$

$$\text{Loss in amplitude/cycle} = \frac{4F_d}{k} = \frac{4 \times 10}{10^4} \text{ rad}$$

$$= 0.004 \text{ rad.}$$

Number of cycles for motion to cease

$$= \frac{0.05}{0.004} = 12\tfrac{1}{2}.$$

The beam is in the initial (equilibrium) position when motion ceases. The motion is shown below.

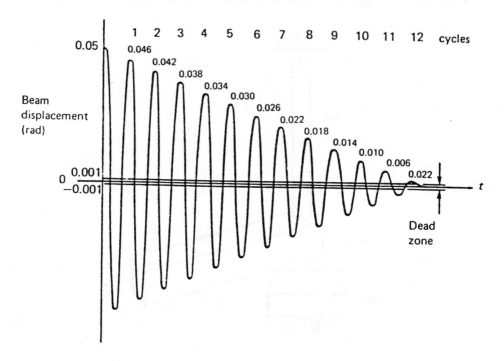

2.2.3 Vibration with combined viscous and Coulomb damping

The free vibration of dynamic structures with viscous damping is characterized by an exponential decay of the oscillation, whereas structures with Coulomb damping possess a linear decay of oscillation. Many real structures have both forms of damping, so that their vibration decay is a combination of exponential and linear functions.

The two damping actions are sometimes amplitude-dependent, so that initially the decay is exponential, say, and only towards the end of the oscillation does the Coulomb effect show. In the analyses of these cases the Coulomb effect can easily be separated from the total damping to leave the viscous damping alone. The exponential decay with viscous damping can be checked by plotting the amplitudes on logarithmic–linear axes when the decay should be seen to be linear.

If the Coulomb and viscous effects cannot be separated in this way, a mixture of linear and exponential decay functions have to be found by trial and error in order to conform to the experimental data.

2.2.4 Vibration with hysteretic damping

Experiments on the damping that occurs in solid materials and structures that have been subjected to cyclic stressing have shown the damping force to be independent of frequency. This internal, or material, damping is referred to as hysteretic damping. Since the viscous damping force $c\dot{x}$ is dependent upon the frequency of oscillation, it is not a suitable way of modelling the internal damping of solids and structures. The analysis of systems and structures with this form of damping therefore requires the damping force $c\dot{x}$ to be divided by the frequency of oscillation ω. Thus the equation of motion becomes $m\ddot{x} + (c/\omega)\dot{x} + kx = 0$.

However, it has been observed from experiments carried out on many materials and structures that under harmonic forcing the stress leads the strain by a constant angle, α.

Thus for an harmonic strain, $\varepsilon = \varepsilon_0 \sin vt$, where v is the forcing frequency, the induced stress is $\sigma = \sigma_0 \sin (vt + \alpha)$. Hence

$$\sigma = \sigma_0 \cos \alpha \sin vt + \sigma_0 \sin \alpha \cos vt$$

$$= \sigma_0 \cos \alpha \sin vt + \sigma_0 \sin \alpha \sin \left(vt + \frac{\pi}{2} \right).$$

The first component of stress is in phase with the strain ε, whilst the second component is in quadrature with ε and $\pi/2$ ahead. Putting $j = \sqrt{(-1)}$,

$$\sigma = \sigma_0 \cos \alpha \sin vt + j\sigma_0 \sin \alpha \sin vt.$$

Hence a complex modulus E^* can be formulated, where

$$E^* = \frac{\sigma}{\varepsilon} = \frac{\sigma_0}{\varepsilon_0} \cos \alpha + j \frac{\sigma_0}{\varepsilon_0} \sin \alpha$$

$$= E' + jE'',$$

where E' is the in-phase or storage modulus, and E'' is the quadrature or loss modulus.

The loss factor η, which is a measure of the hysteretic damping in a structure, is equal to E''/E', that is, $\tan \alpha$.

It is not usually possible to separate the stiffness of a structure from its hysteretic damping, so that in a mathematical model these quantities have to be considered together. The complex stiffness k^* is given by $k^* = k(1 + j\eta)$, where k is the static stiffness and η is the hysteretic damping loss factor.

2.2.5 Complex stiffness

In most real structures it is not possible to separate the stiffness and damping effects because they are inherent properties which are often coupled. Realistic mathematical models of structures therefore require these quantities to be considered together in the form of a complex stiffness. Although this is rather an awkward physical concept it is widely used in analysis.

The complex stiffness k^* is equal to $k(1 + j\eta)$, where k is the static stiffness, $j = \sqrt{-1}$ and η is the hysteretic damping loss factor.

The equation of free motion for a single degree of freedom system with hysteretic damping is therefore $m\ddot{x} + k^*x = 0$. Fig. 2.19 shows a single degree of freedom model with hysteretic damping of coefficient c_H.

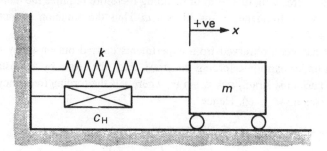

Fig. 2.19.   Single degree of freedom with hysteretic damping.

The equation of motion is

$$m\ddot{x} + (c_H/\omega)\dot{x} + kx = 0.$$

Now if $x = Xe^{j\omega t}$,

$$\dot{x} = j\omega x \quad \text{and} \quad \left(\frac{c_H}{\omega}\right)\dot{x} = jc_H x.$$

Thus the equation of motion becomes

$$m\ddot{x} + (k + jc_H)x = 0.$$

Since

$$k + jc_H = k\left(1 + \frac{jc_H}{k}\right) = k(1 + j\eta) = k^*,$$

we can write

$$m\ddot{x} + k^*x = 0,$$

that is, the combined effect of the elastic and hysteretic resistance to motion can be represented as a complex stiffness, k^*.

A range of values of η for some common engineering materials is given in the following table. For more detailed information on material damping mechanisms and loss factors, see *Damping of Materials and Members in Structural Mechanics* by B. J. Lazan (Pergamon, 1968), and Chapter 5.

Material	Loss factor
Aluminium–pure	0.00002–0.002
Aluminium alloy–dural	0.0004–0.001
Steel	0.001–0.008
Lead	0.008–0.014
Cast iron	0.003–0.03
Manganese copper alloy	0.05–0.1
Rubber–natural	0.1–0.3
Rubber–hard	1.0
Glass	0.0006–0.002
Concrete	0.01–0.06

2.2.6 Energy dissipated by damping

The energy dissipated per cycle by the viscous damping force in a single degree of freedom vibrating system is approximately

$$4 \int_0^x c\dot{x} \, dx,$$

if $x = X \sin \omega t$ is assumed for the complete cycle. The energy dissipated is therefore

$$4 \int_0^{\pi/2} cX^2\omega^2 \cos^2 \omega t \, dt = \pi c \omega X^2.$$

The energy dissipated per cycle by Coulomb damping is $4F_d X$ approximately. Thus an equivalent viscous damping coefficient for Coulomb damping c_d can be deduced, where

$$\pi c_d \omega X^2 = 4F_d X,$$

that is,

$$c_d = \frac{4F_d}{\pi \omega X}.$$

The energy dissipated per cycle by a force F acting on a system with hysteretic damping is $\int F \, dx$, where $F = k^*x = k(1 + j\eta)x$, and x is the displacement.

For harmonic motion $x = X \sin \omega t$,

so

$$F = kX \sin \omega t + j\eta kX \sin \omega t$$
$$= kX \sin \omega t + \eta kX \cos \omega t.$$

Now

$$\sin \omega t = \frac{x}{X}, \quad \text{therefore} \quad \cos \omega t = \frac{\sqrt{(X^2 - x^2)}}{X}.$$

Thus

$$F = kx \pm \eta k\sqrt{(X^2 - x^2)}.$$

This is the equation of an ellipse as shown in Fig. 2.20. The energy dissipated is given by the area enclosed by the ellipse.

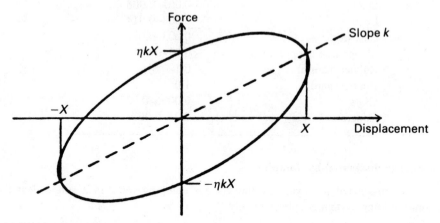

Fig. 2.20.   Elliptical force–displacement relationship for a system with hysteretic damping.

Hence

$$\int F dx = \int_0^x (kx \pm \eta k\sqrt{(X^2 - x^2)})dx$$

$$= \pi X^2 \eta k.$$

An equivalent viscous damping coefficient c_H is given by

$$\pi c_H \omega X^2 = \pi \eta k X^2,$$

that is

$$c_H = \frac{\eta k}{\omega},$$

Note also that $c = c_H \omega.$

Example 11

A single degree of freedom system has viscous damping, with $\zeta = 0.02$. Find the energy dissipated per cycle as a function of the energy in the system at the start of that cycle. Also find the amplitude of the 12th cycle if the amplitude of the 3rd cycle is 1.8 mm.

$\zeta \ll 1$, so ln $(X_1/X_2) = 2\pi\zeta = 0.126.$

Thus

$$X_1/X_2 = e^{0.126} = 1.134.$$

Energy at start of cycle $= \frac{1}{2}kX_1^2$ (stored as strain energy in spring)

Energy at end of cycle $= \frac{1}{2}kX_2^2$

$$\frac{\text{Energy dissipated during cycle}}{\text{Energy at start of cycle}} = \frac{\frac{1}{2}kX_1^2 - \frac{1}{2}kX_2^2}{\frac{1}{2}kX_1^2} = 1 - (X_2/X_1)^2 = 1 - 0.7773 = 0.223,$$

that is, 22.3% of the initial energy is dissipated in one cycle.

Now

$$X_1/X_2 = 1.134, \quad X_2/X_3 = 1.134, \quad \ldots , \quad (X_{n-1})/X_n = 1.134.$$

therefore

$$X_3/X_{12} = (1.134)^9 = 3.107$$

that is

$$X_{12} = \frac{1.8}{3.107} = 0.579 \text{ mm.}$$

2.3   FORCED VIBRATION

Many real structures are subjected to periodic excitation. This may be due to unbalanced rotating or reciprocating components of machinery or equipment, wind or current effects, or a shaking foundation. Usually very low vibration amplitudes are required over a large range of exciting forces and frequencies to keep dynamic stresses, noise, fatigue and other effects to acceptable levels. Some periodic forces are harmonic, but even if they are not, they can be represented as a series of harmonic functions using Fourier analysis techniques. Because of this the response of structures and dynamic systems subjected to harmonic exciting forces and motions must be studied. Non-periodic excitation such as shock, impulse and random, are considered later, in sections 2.3.5 to 2.3.9.

2.3.1   Response of a viscous damped structure to a simple harmonic exciting force with constant amplitude

In the system shown in Fig. 2.21, the body of mass m is connected by a spring and viscous damper to a fixed support, whilst a harmonic force of circular frequency v and amplitude F acts upon it, in the line of motion.

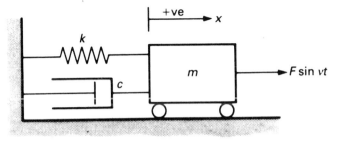

Fig. 2.21.   Single degree of freedom model of a forced system with viscous damping.

The equation of motion is

$$m\ddot{x} + c\dot{x} + kx = F \sin vt. \tag{2.11}$$

The solution to $m\ddot{x} + c\dot{x} + kx = 0$, which has already been studied, is the complementary function; it represents the initial vibration which quickly dies away. The sustained motion is given by the particular solution. A solution $x = X (\sin vt - \phi)$ can be assumed, because this represents simple harmonic motion at the frequency of the exciting force with a displacement vector which lags the force vector by ϕ, that is, the motion occurs after the application of the force.

Assuming $x = X \sin(vt - \phi)$,

$$\dot{x} = Xv \cos (vt - \phi) = Xv \sin (vt - \phi + \tfrac{1}{2}\pi),$$

and

$$\ddot{x} = - Xv^2 \sin (vt - \phi) = Xv^2 \sin (vt - \phi + \pi).$$

The equation of motion (2.11) thus becomes

$$mXv^2 \sin (vt - \phi + \pi) + cXv \sin (vt - \phi + \pi/2) + kX \sin (vt - \phi)$$
$$= F \sin vt.$$

A vector diagram of these forces can now be drawn (Fig. 2.22).

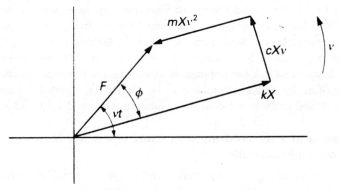

Fig. 2.22.   Force vector diagram.

From the diagram,

$$F^2 = (kX - mXv^2)^2 + (cXv)^2,$$

or

$$X = F/\sqrt{((k - mv^2)^2 + (cv)^2)}, \tag{2.12}$$

and

$$\tan \phi = cXv/(kX - mXv^2).$$

Thus the steady-state solution to equation (2.11) is

$$x = \frac{F}{\sqrt{((k - mv^2)^2 + (cv)^2)}} \sin (vt - \phi),$$

where

$$\phi = \tan^{-1}\left(\frac{cv}{k - mv^2}\right).$$

The complete solution includes the transient motion given by the complementary function:

$$x = Ae^{-\zeta\omega t} \sin(\omega\sqrt{(1 - \zeta^2)}t + \alpha).$$

Fig. 2.23 shows the combined motion.

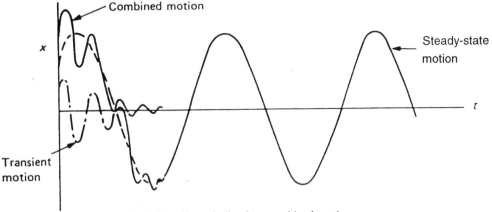

Fig. 2.23.   Forced vibration, combined motion.

Equation (2.12) can be written in a more convenient form if we put

$$\omega = \sqrt{\left(\frac{k}{m}\right)} \text{ rad/s } \text{ and } X_s = \frac{F}{k}.$$

Then

$$\frac{X}{X_s} = \frac{1}{\sqrt{\left\{\left[1 - \left(\frac{v}{\omega}\right)^2\right]^2 + \left[2\zeta\frac{v}{\omega}\right]^2\right\}}},$$ (2.13)

and $\phi = \tan^{-1}\left[\dfrac{2\zeta(v/\omega)}{1 - (v/\omega)^2}\right]$

X/X_s is known as the dynamic magnification factor, because X_s is the static deflection of the system under a steady force F, and X is the dynamic amplitude.

By considering different values of the frequency ratio v/ω, we can plot X/X_s and ϕ as functions of frequency for various values of ζ. Figs 2.24 and 2.25 show the results.

The effect of the frequency ratio on the force vector diagram is shown in Fig. 2.26.

The importance of mechanical vibration arises mainly from the large values of X/X_s experienced in practice when v/ω has a value near unity: this means that a small harmonic

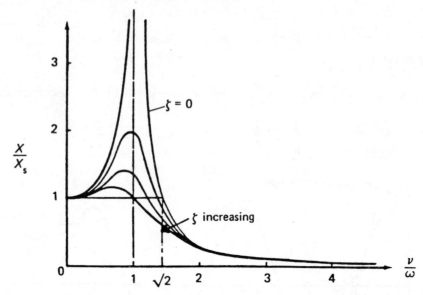

Fig. 2.24.   Amplitude–frequency response for system of Fig. 2.21.

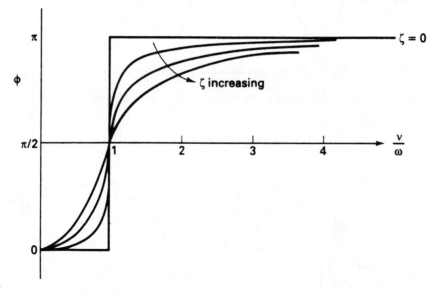

Fig. 2.25.   Phase–frequency response for system of Fig. 2.21.

force can produce a large amplitude of vibration. The phenomenon known as *resonance* occurs when the forcing frequency is equal to the natural frequency, that is when $v/\omega = 1$. The maximum value of X/X_s actually occurs at values of v/ω less than unity: the value can be found by differentiating equation (2.13) with respect to v/ω. Hence

$$(v/\omega)_{(X/X_s)\text{max}} = \sqrt{(1 - 2\zeta^2)} \simeq 1 \text{ for } \zeta \text{ small,}$$

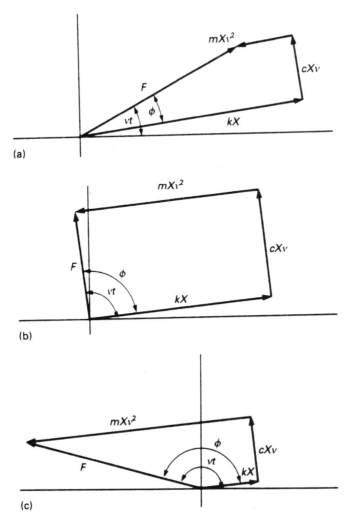

Fig. 2.26.   Forced vibration vector diagrams: (a) $v/\omega \ll 1$, exciting force approximately equal to spring force; (b) $v/\omega = 1$, exciting force equal to damping force, and inertia force equal to spring force; (c) $v/\omega \gg 1$, exciting force nearly equal to inertia force.

and

$$(X/X_s)_{max} = 1/(2\zeta\sqrt{(1 - \zeta^2)}).$$

For small values of ζ, $(X/X_s)_{max} \simeq 1/2\zeta$ which is the value pertaining to $v/\omega = 1$; $1/2\zeta$ is a measure of the damping in a system and is referred to as the Q factor.

Both *reciprocating* and *rotating unbalance* in a system produce an exciting force of the inertia type and result in the amplitude of the exciting force being proportional to the square of the frequency of excitation.

For an unbalanced body of mass m_r at an effective radius r, rotating at an angular speed v, the exciting force is therefore $m_r r v^2$. If this force is applied to a single degree of freedom system such as that in Fig. 2.21, the component of the force in the direction of motion is $m_r r v^2 \sin vt$, and the amplitude of vibration is

$$X = \frac{(m_r/m)r(v/\omega)^2}{\sqrt{((1 - (v/\omega)^2)^2 + (2\zeta\, v/\omega)^2)}}.$$ (2.14)

(see equation (2.13)).

The value of v/ω for maximum X found by differentiating equation (2.14) is given by

$$(v/\omega)_{X\max} = 1/\sqrt{(1 - 2\zeta^2)}$$

that is, the peak of the response curve occurs when $v > \omega$. This is shown in Fig. 2.27. Also,

$$X_{\max} = (m_r/m)r/2\zeta\sqrt{(1 - \zeta^2)}.$$

It can be seen that away from the resonance condition ($v/\omega = 1$) the system response is not greatly affected by damping unless this happens to be large. Since in most mechanical systems the damping is small ($\zeta < 0.1$) it is often permissible to neglect the damping when evaluating the frequency for maximum amplitude and also the amplitude–frequency response away from the resonance condition.

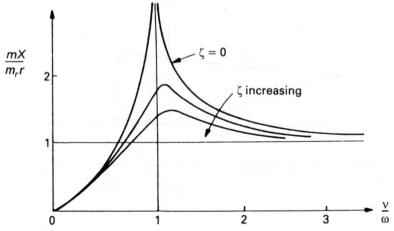

Fig. 2.27.   Amplitude–frequency response, with rotating unbalance excitation.

It can be seen from Figs 2.24, 2.26 and 2.27 that the system response at low frequencies ($\ll\omega$) is stiffness-dependent, and that in the region of resonance the response is damping-dependent, whereas at high frequencies ($\gg\omega$) the response is governed by the system mass. It is most important to realize this when attempting to reduce the vibration of a structure. For example, the application of increased damping will have little effect if the excitation and response frequencies are in a region well away from resonance, such as that controlled by the mass of the structure.

2.3.2   Response of a viscous damped structure supported on a foundation subjected to harmonic vibration

The system considered is shown in Fig. 2.28. The foundation is subjected to harmonic vibration $A \sin vt$ and it is required to determine the response, x, of the body.

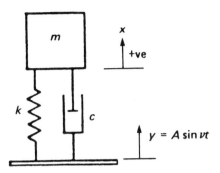

Fig. 2.28.   Single degree of freedom model of a vibrated system with viscous damping.

The equation of motion is

$$m\ddot{x} = c(\dot{y} - \dot{x}) + k(y - x). \tag{2.15}$$

If the displacement of the body relative to the foundation, u, is required, we may write $u = x - y$, and equation (2.15) becomes

$$m\ddot{u} + c\dot{u} + ku = -m\ddot{y} = mv^2 A \sin vt.$$

This equation is similar to (2.11) so that the solution may be written directly as

$$u = \frac{A(v/\omega)^2}{\sqrt{((1 - (v/\omega)^2)^2 + (2\zeta v/\omega)^2)}} \sin\left(vt - \tan^{-1}\frac{2\zeta(v/\omega)}{1 - (v/\omega)^2}\right).$$

If the absolute motion of the body is required we rewrite equation (2.15) as

$$m\ddot{x} + c\dot{x} + kx = c\dot{y} + ky$$
$$= cAv \cos vt + kA \sin vt$$
$$= A\sqrt{(k^2 + (cv)^2)} \sin (vt + \alpha)$$

where

$$\alpha = \tan^{-1}\frac{cv}{k}.$$

Hence, from the previous result,

$$x = \frac{A\sqrt{(k^2 + (cv)^2)}}{\sqrt{((k - mv^2)^2 + (cv)^2)}} \sin (vt - \phi + \alpha).$$

The motion transmissibility is defined as the ratio of the amplitude of the absolute body vibration to the amplitude of the foundation vibration. Thus,

$$\text{motion transmissibility} = \frac{X}{A}$$

$$= \frac{\sqrt{\left[1 + \left(2\zeta\frac{v}{\omega}\right)^2\right]}}{\sqrt{\left\{\left[1 - \left(\frac{v}{\omega}\right)^2\right]^2 + \left[2\zeta\frac{v}{\omega}\right]^2\right\}}}.$$

2.3.2.1 Vibration isolation

The dynamic forces produced by machinery are often very large. However, the force transmitted to the foundation or supporting structure can be reduced by using flexible mountings with the correct properties; alternatively a machine can be isolated from foundation vibration by using the correct flexible mountings. Fig. 2.29 shows a model of such a system.

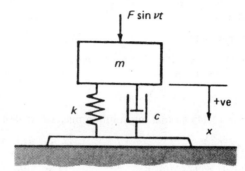

Fig. 2.29.   Single degree of freedom system with foundation.

The force transmitted to the foundation is the sum of the spring force and the damper force. Thus the transmitted force $= kx + c\dot{x}$ and F_T, the amplitude of the transmitted force is given by

$$F_T = \sqrt{[(kX)^2 + (cvX)^2]}.$$

The force transmission ratio or transmissibility, T_R, is given by

$$T_R = \frac{F_T}{F} = \frac{X\sqrt{[k^2 + (cv)^2]}}{F}$$

since

$$X = \frac{F/k}{\sqrt{\left\{\left[1 - \left(\frac{v}{\omega}\right)^2\right]^2 + \left[2\zeta\frac{v}{\omega}\right]^2\right\}}},$$

$$T_R = \frac{\sqrt{\left[1 + \left(2\zeta\frac{v}{\omega}\right)^2\right]}}{\sqrt{\left\{\left[1 - \left(\frac{v}{\omega}\right)^2\right]^2 + \left[2\zeta\frac{v}{\omega}\right]^2\right\}}}.$$

Therefore the force and motion transmissibilities are the same.

The effect of v/ω on T_R is shown in Fig. 2.30. It can be seen that for good isolation $v/\omega > \sqrt{2}$, hence a low value of ω is required which implies a low stiffness, that is, a flexible mounting: this may not always be acceptable in practice where a certain minimum stiffness is usually necessary to satisfy operating criteria.

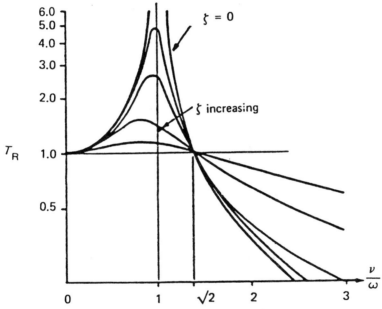

Fig. 2.30.   Transmissibility–frequency ratio response.

It is particularly important to be able to isolate vibration sources because structure-borne vibration can otherwise be easily transmitted to parts that radiate well, and serious noise problems can occur. Theoretically, low stiffness isolators are desirable to give a low natural frequency. However, this often results in isolators that are too soft and stability problems may arise. The system can be attached rigidly to a large block which effectively increases its mass so that stiffer isolators can be used. The centre of mass of the combined system is also lowered, giving improved stability. For the best response a mounting system may be designed with snubbers, which control the large amplitudes while providing little or no damping when the amplitudes are small.

There are four types of spring material commonly used for resilient mountings and vibration isolation: air, metal, rubber and cork. Air springs can be used for very low-frequency suspensions: resonance frequencies as low as 1 Hz can be achieved whereas metal springs can only be used for resonance frequencies greater than about 1.3 Hz. Metal springs can transmit high frequencies, however, so rubber or felt pads are often used to prohibit metal-to-metal contact between the spring and the structure. Different forms of spring element can be used such as coil, torsion, cantilever and beam. Rubber can be used in shear or compression but rarely in tension. It is important to determine the dynamic stiffness of a rubber isolator because this is generally much greater than the static stiffness. Also rubber possesses some inherent damping although this may be sensitive to amplitude, frequency and temperature. Natural frequencies from 5 Hz upwards can be achieved. Cork is one of the oldest materials used for vibration isolation. It is usually used in compression and natural frequencies of 25 Hz upwards are typical.

Example 12

A spring-mounted body moves with velocity v along an undulating surface, as shown.

The body has a mass m and is connected to the wheel by a spring of stiffness k, and a viscous damper whose damping coefficient is c. The undulating surface has a wavelength L and an amplitude h.

Derive an expression for the ratio of amplitudes of the absolute vertical displacement of the body to the surface undulations.

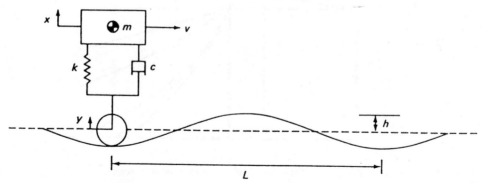

The system can be considered as

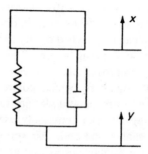

where

$$y = h \cos \frac{2\pi z}{L} \text{ and } z = vt.$$

Hence

$$y = h \cos \frac{2\pi v}{L} t = h \cos vt, \text{ where } v = \frac{2\pi v}{L}.$$

The FBDs are

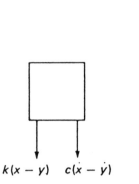
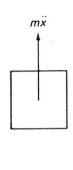

$$k(x - y) \quad c(\dot{x} - \dot{y})$$

Hence the equation of motion is

$$m\ddot{x} = -k(x - y) - c(\dot{x} - \dot{y}),$$

or

$$m\ddot{x} + c\dot{x} + kx = c\dot{y} + ky.$$

Now

$$y = h \cos vt,$$

so

$$m\ddot{x} + c\dot{x} + kx = \sqrt{[k^2 + (cv)^2]} h \sin (vt + \phi).$$

Hence, if $x = X_0 \sin (vt + \alpha)$, then

$$X_0 = \frac{h\sqrt{[k^2 + (cv)^2]}}{\sqrt{[(k - mv^2)^2 + (cv)^2]}}$$

So,

$$\frac{X_0}{h} = \frac{\sqrt{\left[k^2 + \left(\dfrac{2\pi v}{L} c\right)^2\right]}}{\sqrt{\left\{\left[k - \left(\dfrac{2\pi v}{L}\right)^2 m\right]^2 + \left(\dfrac{2\pi v}{L} c\right)^2\right\}}}$$

Example 13

The vibration on the floor in a building is SHM at a frequency in the range 15–60 Hz. It is desired to install sensitive equipment in the building which must be insulated from floor vibration. The equipment is fastened to a small platform which is supported by three similar springs resting on the floor, each carrying an equal load. Only vertical motion occurs. The combined mass of the equipment and platform is 40 kg, and the equivalent viscous damping ratio of the suspension is 0.2.

Find the maximum value for the spring stiffness, if the amplitude of transmitted vibration is to be less than 10% of the floor vibration over the given frequency range.

$$T_R = \frac{\sqrt{\left[1 + \left(2\zeta\frac{v}{\omega}\right)^2\right]}}{\sqrt{\left\{\left[1 - \left(\frac{v}{\omega}\right)^2\right]^2 + \left[2\zeta\frac{v}{\omega}\right]^2\right\}}}$$

$T_R = 0.1$ with $\zeta = 0.2$ is required, thus

$$\left[1 - \left(\frac{v}{\omega}\right)^2\right]^2 + \left[0.4\left(\frac{v}{\omega}\right)\right]^2 = 100\left[1 + \left(0.4\frac{v}{\omega}\right)^2\right],$$

that is,

$$\left(\frac{v}{\omega}\right)^4 - 17.84\left(\frac{v}{\omega}\right)^2 - 99 = 0.$$

Hence

$$\frac{v}{\omega} = 4.72.$$

When

$$v = 15 \times 2\pi \text{ rad/s}, \ \omega = 19.97 \text{ rad/s}.$$

Since

$$\omega = \sqrt{\left(\frac{k}{m}\right)} \quad \text{and} \quad m = 40 \text{ kg},$$

total $k = 15\,935$ N/m,

that is, the stiffness of each spring $= 15\,935/3$ N/m $= 5.3$ kN/m.
   The amplitude of the transmitted vibration will be less than 10% at frequencies above 15 Hz.

Example 14

A machine of mass m generates a disturbing force $F \sin vt$; to reduce the force transmitted to the supporting structure, the machine is mounted on a spring of stiffness k with a damper in parallel. Compare the effectiveness of this isolation system for viscous and hysteretic damping.

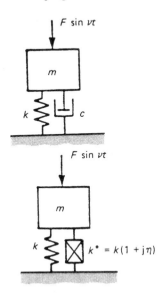

Viscous damping. From section 2.3.2.1,

$$T_R = \frac{F_T}{F} = \frac{\sqrt{\left[1 + \left(2\zeta\frac{v}{\omega}\right)^2\right]}}{\sqrt{\left\{\left[1 - \left(\frac{v}{\omega}\right)^2\right]^2 + \left[2\zeta\frac{v}{\omega}\right]^2\right\}}}$$

Hysteretic damping. From section 2.2.6,

Putting $\eta = \dfrac{cv}{k} = 2\zeta\dfrac{v}{\omega}$,

$$T_R = \frac{F_T}{F} = \frac{\sqrt{(1 + \eta^2)}}{\sqrt{\left\{\left[1 - \left(\frac{v}{\omega}\right)^2\right]^2 + \eta^2\right\}}}.$$

The effectiveness of these isolators can be compared using these expressions for T_R. The results are given in the table below.

It can be seen that the isolation effects are similar for the viscous and hysteretically damped isolators, except at high frequency ratios when the hysteretic damping gives much better attentuation of T_R. At these frequencies it is better to decouple the viscous damped isolator by attaching small springs or rubber bushes at each end.

	Viscously damped isolator		Hysteretically damped isolator	
Value of T_R when $v = 0^+$	1		1	
Frequency ratio v/ω for resonance	1		1	
Value of T_R at resonance	$\dfrac{\sqrt{[1 + (2\zeta)^2]}}{2\zeta}$	$\simeq \dfrac{1}{2\zeta}$	$\dfrac{\sqrt{(1 + \eta^2)}}{\eta}$	$\simeq \dfrac{1}{\eta}$
Value of T_R when $v/\omega = \sqrt{2}$	1		1	
Frequency ratio v/ω for isolation	$> \sqrt{2}$		$> \sqrt{2}$	
High frequency, $v/\omega \gg 1$, attenuation of T_R	$\dfrac{2\zeta}{v/\omega}$		$\dfrac{1}{(v/\omega)^2}$	

Example 15

A motor-generator set of mass 100 kg is installed using antivibration (AV) mountings which deflect 1 mm under the static weight of the set. The mountings are effectively undamped and from dynamic test results it is found that the static stiffness and the dynamic stiffness are the same.

When running at 1480 rev/min the amplitude of vibration of the set is measured to be 0.2 mm To reduce this vibration, it is proposed to fasten the motor generator to a concrete block of mass 300 kg which is then to be mounted on the same AV mounts as before. Calculate the new amplitude of vibration.

For undamped mounts,

$$x = \frac{F}{k - mv^2}.$$

Initially,

$$x_1 = \frac{F}{k - m_1 v^2},$$

and when on the block,

$$x_2 = \frac{F}{k - m_2 v^2},$$

that is,

$$\frac{x_1}{x_2} = \frac{k - m_2 v^2}{k - m_1 v^2}.$$

Now

$$k = \frac{100 \times 9.81}{10^{-3}} = 981 \times 10^3 \text{ N/m},$$

$$v = \frac{2\pi \times 1480}{60} = 155 \text{ rad/s},$$

$$m_1 = 100 \text{ kg and } m_2 = 400 \text{ kg}.$$

Thus

$$\frac{x_1}{x_2} = \frac{981 \times 10^3 - 400 \,(155)^2}{981 \times 10^3 - 100 \,(155)^2}$$

$$= \frac{981 - 9600}{981 - 2400} = 6.07.$$

Since

$$x_1 = 0.2 \text{ mm}, \ x_2 = \frac{0.2}{6.07} = 0.033 \text{ mm}.$$

Example 16

A machine of mass 550 kg is flexibly supported on rubber mountings which provide a force proportional to displacement of 210 kN/m, together with a viscous damping force. The machine gives an exciting force of the form $Rv^2 \cos vt$, where R is a constant. At very high speeds of rotation, the measured amplitude of vibration is 0.25 mm, and the maximum amplitude recorded as the speed is slowly increased from zero is 2 mm. Find the value of R and the damping ratio.

Now,

$$X = Rv^2/(\sqrt{(k - mv^2)^2 + c^2v^2}).$$

If v is large,

$$X \to Rv^2/(\sqrt{(m^2v^4)}) = R/m.$$

Hence

$$R = mX = (550 \times 0.25)/1000 = 0.1375 \text{ kg m}.$$

For maximum X, $dX/dv = 0$, hence $v^2 = 2k^2/(2mk - c^2)$, and

$$X_{\text{max}} = \frac{R/2m}{\zeta\sqrt{(1 - \zeta^2)}}.$$

So

$$\zeta\sqrt{(1 - \zeta^2)} = R/(2mX_{\text{max}}) = 0.1375/(2 \times 550 \times 2 \times 10^{-3}) = 0.0625,$$

that is,

$$\zeta = 0.0625.$$

2.3.3 Response of a Coulomb damped structure to a simple harmonic exciting force with constant amplitude

In the system shown in Fig. 2.31 the damper relies upon dry friction.

The equation of motion is non-linear because the constant friction force F_d always opposes the motion:

$$m\ddot{x} + kx + F_d = F \sin vt.$$

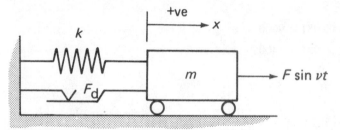

Fig. 2.31.   Single degree of freedom model of a forced system with Coulomb damping.

If F_d is large compared to F, discontinuous motion will occur, but in most structures F_d is usually small so that an approximate continuous solution is valid. The approximate solution is obtained by linearizing the equation of motion; this can be done by expressing F_d in terms of an equivalent viscous damping coefficient, c_d. From section 2.2.6,

$$c_d = \frac{4F_d}{\pi v X}.$$

The solution to the linearized equation of motion gives the amplitude X of the motion as

$$X = \frac{F}{\sqrt{[(k - mv^2)^2 + (c_d v)^2]}}.$$

Hence

$$X = \frac{F}{\sqrt{[(k - mv^2)^2 + (4F_d/\pi X)^2]}},$$

that is,

$$\frac{X}{X_s} = \frac{\sqrt{(1 - (4F_d/\pi F)^2)}}{1 - (v/\omega)^2}.$$

This expression is satisfactory for small damping forces, but breaks down if $4F_d/\pi F < 1$; that is, $F_d > (\pi/4)F$.

At resonance the amplitude is not limited by Coulomb friction.

2.3.4 Response of a hysteretically damped structure to a simple harmonic exciting force with constant amplitude

In the single degree of freedom model shown in Fig. 2.32 the damping is hysteretic. The equation of motion is

$$m\ddot{x} + k^*x = F \sin vt.$$

Since

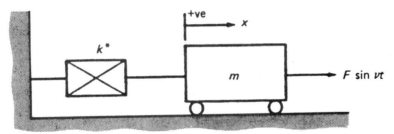

Fig. 2.32.   Single degree of freedom model of a forced system with hysteretic damping.

$$k^* = k(1 + j\eta),$$

$$x = \frac{F \sin vt}{(k - mv^2) + j\eta k}$$

and

$$\frac{X}{X_s} = \frac{1}{\sqrt{([1 - (v/\omega)^2]^2 + \eta^2)}}.$$

This result can also be obtained from the analysis of a viscous damped system by substituting $c = \eta k/v$.
    It should be noted that if $c = \eta k/v$, at resonance $c = \eta\sqrt{(km)}$; that is,
$\eta = 2\zeta = 1/Q$.

Since $Q = \dfrac{1}{\eta}$,

if a structure is made from a concrete material for which $\eta = 0.02$, a Q factor of 50 may be expected. For a steel, with $\eta = 0.005$ a Q factor of 200 may be expected and for a cast iron with $\eta = 0.01$, $Q = 100$. In practice Q values very much lower than these occur, often by an order of magnitude; that is, Q factors of 10 or less are common. Most of the additional damping found in structures originates in the joints between the connected components of the structure. Joint damping is often the most significant form of damping in a structure and keeps the dynamic response to acceptable levels; it is fully discussed in Chapter 5.

2.3.5 Response of a structure to a suddenly applied force

Consider a single degree of freedom undamped system, such as the system shown in Fig. 2.33, which has been subjected to a suddenly applied force, F. The equation of motion is $m\ddot{x} + kx = F$. The solution to this equation comprises a complementary function $A \sin \omega t + B \cos \omega t$, where $\omega = \sqrt{(k/m)}$ rad/s together with a particular solution. The particular solution may be found by using the D-operator. Thus the equation of motion can

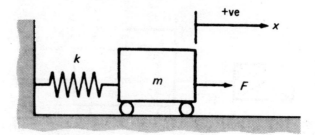

Fig. 2.33.   Single degree of freedom model with constant exciting force.

be written

$$\left(1 + \frac{D^2}{\omega^2}\right)x = \frac{F}{k},$$

and

$$x = \left(1 + \frac{D^2}{\omega^2}\right)^{-1}\frac{F}{k} = \left(1 - \frac{D^2}{\omega^2} - \dots\right)\frac{F}{k} = \frac{F}{k};$$

that is, the complete solution to the equation of motion is

$$x = A \sin \omega t + B \cos \omega t + \frac{F}{k}.$$

If the initial conditions are such that $x = \dot{x} = 0$ at $t = 0$, then $B = -F/k$ and $A = 0$. Hence

$$x = \frac{F}{k}(1 - \cos \omega t).$$

The motion is shown in Fig. 2.34. It will be seen that the maximum dynamic displacement is twice the static displacement occurring under the same load. This is an important consideration in structures subjected to suddenly applied loads.

If the structure possesses viscous damping of coefficient c, the solution to the equation of motion is $x = Xe^{-\zeta\omega t}\sin(\omega_v t + \phi) + F/k$.

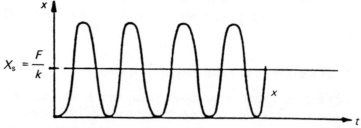

Fig. 2.34.   Displacement–time response for the system shown in Fig. 2.33.

With the same initial conditions as above,

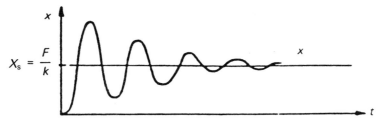

This reduces to the undamped case if $\zeta = 0$. The response of the damped system is shown in Fig. 2.35.

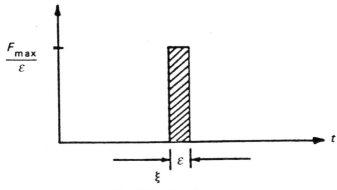

Fig. 2.35.   Displacement–time response for a single degree of freedom system with viscous damping.

2.3.6 Shock excitation

Some structures are subjected to shock or impulse loads arising from suddenly applied, non-periodic, short-duration exciting forces.

The impulsive force shown in Fig. 2.36 consists of a force of magnitude F_{max}/ε which has a time duration of ε.

Fig. 2.36.   Impulse.

The impulse is equal to

$$\int_{t}^{t+\varepsilon} \left(\frac{F_{max}}{\varepsilon} \right) dt.$$

When F_{max} is equal to unity, the force in the limiting case $\varepsilon \rightarrow 0$ is called either the unit impulse or the delta function, and is identified by the symbol $\delta(t - \xi)$, where

$$\int_{0}^{\infty} \delta(t - \xi) d\xi = 1.$$

Since $F\, dt = m\, dv$, the impulse F_{max} acting on a body will result in a sudden change in its velocity without an appreciable change in its displacement. Thus the motion of a single degree of freedom system excited by an impulse F_{max} corresponds to free vibration with initial conditions $x = 0$ and $\dot{x} = v_0 = F_{max}/m$ at $t = 0$.

Once the response $g(t)$, say, to a unit impulse excitation is known, it is possible to establish the equation for the response of a system to an arbitrary exciting force $F(t)$. For this the arbitrary pulse is considered to comprise a series of impulses as shown in Fig. 2.37.

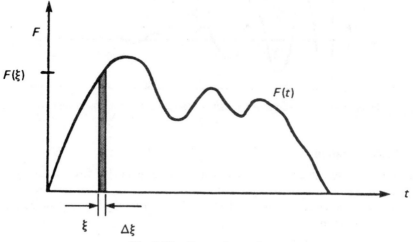

Fig. 2.37. Force–time pulse.

If one of the impulses is examined which starts at time ξ, its magnitude is $F(\xi)\delta\xi$, and its contribution to the system response at time t is found by replacing the time with the elapsed time $(t - \xi)$ as shown in Fig. 2.38.

If the system can be assumed to be linear, the principle of superposition can be applied, so that

$$x(t) = \int_0^t F(\xi)g(t - \xi)d\xi.$$

This is known as the Duhamel integral.

2.3.7 Wind- or current-excited oscillation

A structure exposed to a fluid stream is subjected to a harmonically varying force in a direction perpendicular to the stream. This is because of eddy, or vortex, shedding on alternate sides of the structure on the leeward side. Tall structures such as masts, bridges and chimneys are susceptible to excitation from steady winds blowing across them. Consider a circular cylinder of diameter D exposed to a fluid which flows past the cylinder with a velocity v. When v is large enough, vortices are formed in the wake which are shed in a regular pattern over a wide range of Reynolds' numbers.

$$\text{Reynolds number} = \frac{vD\rho}{\mu},$$

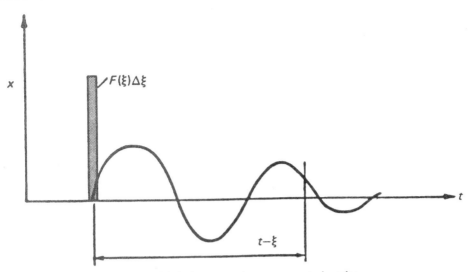

Fig. 2.38.   Displacement–time response to impulse.

where ρ and μ are the mass density and viscosity of the fluid, respectively.

The vortices are shed from opposite sides of the cylinder with a frequency f_s. The Strouhal number relates the excitation frequency, f_s, to the velocity of fluid flow, v (m/s), and the hydraulic mean diameter, D(m), of the structure as follows:

$$\text{Strouhal number} = \frac{f_s D}{v}.$$

This vortex shedding causes an alternating pressure on each side of the cylinder, which acts as a harmonically varying force which is perpendicular to the direction of the undisturbed flow of magnitude

$$\tfrac{1}{2}C_D \rho v^2 A$$

where C_D is the drag coefficient and A is the projected area of the cylinder perpendicular to the direction of flow. If the frequency f_s is close to the natural frequency of the structure, resonance may occur.

For a structure,

$$D = \frac{4 \times \text{area of cross-section}}{\text{circumference}},$$

so that for a chimney of circular cross-section and diameter d,

$$D = \frac{4(\pi/4)\, d^2}{\pi d} = d,$$

and for a building of rectangular cross-section $a \times b$,

$$D = \frac{4ab}{2(a + b)} = \frac{2ab}{(a + b)}.$$

Experimental evidence suggests a value of 0.2–0.24 for the Strouhal number for most flow rates and wind speeds encountered. This value is valid for Reynolds numbers in the range $3 \times 10^5 - 3.5 \times 10^6$.

For a comprehensive discussion of this form of excitation see *Flow Induced Vibration* by R. D. Blevins (Van Nostrand, 1977).

Example 17

For constructing a tanker terminal in a river estuary a number of cylindrical concrete piles were sunk into the river bed and left free-standing. Each pile was 1 m in diameter and protruded 20 m out of the river bed. The density of the concrete was 2400 kg/m^3 and the modulus of elasticity 14×10^6 kN/m^2. Estimate the velocity of the water flowing past a pile which will cause it to vibrate transversely to the direction of the current, assuming a pile to be a cantilever and taking a value for the Strouhal number

$$\frac{f_s D}{v} = 0.22,$$

where f_s is the frequency of flexural vibrations of a pile, D is the diameter and v is the velocity of the current.

Consider the pile to be a cantilever of mass m, diameter D and length l; then the deflection y at a distance x from the root can be taken to be $y = y_l(1 - \cos \pi x/2l)$, where y_l is the deflection at the free end.

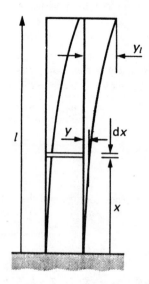

Thus

$$\frac{d^2y}{dx^2} = y_l \left(\frac{\pi}{2l}\right)^2 \cos\left(\frac{\pi x}{2l}\right)$$

$$\int_0^l EI\left(\frac{d^2y}{dx^2}\right)^2 dx = EI\int_0^l y_l^2 \left(\frac{\pi}{2l}\right)^4 \cos^2\left(\frac{\pi x}{2l}\right)dx$$

$$= EIy_l^2 \left(\frac{\pi}{2l}\right)^4 \frac{l}{2}$$

$$\int y^2\, dm = \int_0^l y_l^2 \left(1 - \cos\left(\frac{\pi x}{2l}\right)\right)^2 \left(\frac{m}{l}\right)dx$$

$$= y_l^2 \frac{m}{l}\left(\frac{3}{2} - \frac{4}{\pi}\right)l$$

Hence

$$\omega^2 = \frac{EIy_l^2 \left(\dfrac{\pi}{2l}\right)^4 \dfrac{l}{2}}{y_l^2 \dfrac{m}{l}\left(\dfrac{3}{2} - \dfrac{4}{\pi}\right)l}.$$

Substituting numerical values gives $\omega = 5.53$ rad/s, that is, $f = 0.88$ Hz. When $f_s = 0.88$ Hz resonance occurs; that is, when

$$v = \frac{f_s D}{0.22} = \frac{0.88}{0.22} = 4 \text{ m/s.}$$

2.3.8 Harmonic analysis

A function that is periodic but not harmonic can be represented by the sum of a number of terms, each term representing some multiple of the fundamental frequency. In a *linear* system each of these harmonic terms acts as if it alone were exciting the system, and the system response is the sum of the excitation of all the harmonics.

For example, if the periodic forcing function of a single degree of freedom undamped system is

$$F_1 \sin (vt + \alpha_1) + F_2 \sin (2vt + \alpha_2) + F_3 \sin (3vt + \alpha_3)$$
$$+ \cdots + F_n \sin (nvt + \alpha_n),$$

the steady-state response to $F_1 \sin (vt + \alpha_1)$ is

$$x_1 = \frac{F_1}{k\left(1 - \left(\dfrac{v}{\omega}\right)^2\right)} \sin(vt + \alpha_1),$$

and the response to $F_2 \sin(2vt + \alpha_2)$ is

$$x_2 = \frac{F}{k\left(1 - \left(\dfrac{2v}{\omega}\right)^2\right)} \sin(2vt + \alpha_2),$$

and so on, so that

$$x = \sum_{n=1}^{n} \frac{F_n}{k\left(1 - \left(\dfrac{nv}{\omega}\right)^2\right)} \sin(nvt + \alpha_n).$$

Clearly that harmonic that is closest to the system natural frequency will most influence the response.

A periodic function can be written as the sum of a number of harmonic terms by writing a Fourier series for the function. A Fourier series can be written

$$F(t) = \frac{a_0}{2} + \sum_{n=1}^{\infty} (a_n \cos nvt + b_n \sin nvt),$$

where

$$a_0 = \frac{2}{\tau} \int_0^{\tau} F(t)dt,$$

$$a_n = \frac{2}{\tau} \int_0^{\tau} F(t) \cos vtdt,$$

and

$$b_n = \frac{2}{\tau} \int_0^{\tau} F(t) \sin vtdt.$$

For example, consider the first four terms of the Fourier series representation of the square wave shown in Fig. 2.39 to be required; $\tau = 2\pi$ so $v = 1$ rad/s.

$$F(t) = \frac{a_0}{2} + a_1 \cos vt + a_2 \cos 2vt + \ldots$$

$$+ b_1 \sin vt + b_2 \sin 2vt + \ldots,$$

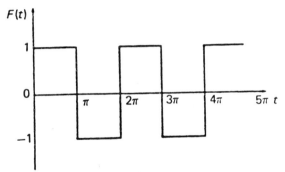

Fig. 2.39.   Square wave.

$$a_0 = \frac{2}{\tau}\int_0^\tau F(t)dt = \frac{2}{2\pi}\int_0^\pi 1\ dt + \frac{2}{2\pi}\int_\pi^{2\pi} -1\ dt = 0,$$

$$a_1 = \frac{2}{\tau}\int_0^\tau F(t)\cos vt dt$$

$$= \frac{2}{2\pi}\int_0^v \cos vt dt + \frac{2}{2\pi}\int_\pi^{2\pi} -\cos vt dt = 0.$$

Similarly

$$a_2 = a_3 = \dots = 0.$$

$$b_1 = \frac{2}{\tau}\int_0^\tau F(t)\sin vt dt$$

$$= \frac{2}{2\pi}\int_0^\pi \sin vt dt + \frac{2}{2\pi}\int_\pi^{2\pi} -\sin vt dt$$

$$= \frac{1}{\pi v}[-\cos vt]_0^\pi + \frac{1}{\pi v}[\cos vt]_\pi^{2\pi} = \frac{4}{\pi v}.$$

Since $v = 1$ rad/s,

$$b_1 = \frac{4}{\pi}.$$

It is found that $b_2 = 0$, $b_3 = 4/3\pi$ and so on. Thus

$$F(t) = \frac{4}{\pi}\left[\sin t + \frac{1}{3}\sin 3t + \frac{1}{5}\sin 5t + \frac{1}{7}\sin 7t + \dots\right]$$

is the series representation of the square wave shown.

If this stimulus is applied to a simple undamped system with $\omega = 4$ rad/s, say, the steady-state response is given by

$$x = \frac{\frac{4}{\pi}\sin t}{1 - \left(\frac{1}{4}\right)^2} + \frac{\frac{4}{3\pi}\sin 3t}{1 - \left(\frac{3}{4}\right)^2} + \frac{\frac{4}{5\pi}\sin 5t}{1 - \left(\frac{5}{4}\right)^2} + \frac{\frac{4}{7\pi}\sin 7t}{1 - \left(\frac{7}{4}\right)^2} \cdots,$$

that is, $x = 1.36 \sin t + 0.97 \sin 3t - 0.45 \sin 5t - 0.09 \sin 7t - \ldots$

Usually three or four terms of the series dominate the predicted response.

It is worth sketching the components of $F(t)$ above to show that they produce a reasonable square wave, whereas the components of x do not. This is an important result.

2.3.9 Random vibration

If the vibration response parameters of a dynamic system are accurately known as functions of time, the vibration is said to be *deterministic*. However, in many systems and processes responses cannot be accurately predicted; these are called *random processes*. Examples of a random process are turbulence, fatigue, the meshing of imperfect gears, surface irregularities, the motion of a car running along a rough road and building vibration excited by an earthquake. Fig. 2.40 shows a random process.

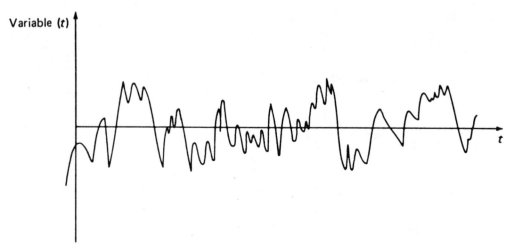

Fig. 2.40.   Example random process variable as a function of t.

A collection of sample functions $x_1(t)$, $x_2(t)$, $x_3(t)$, ..., $x_n(t)$ which make up the random process $x(t)$ is called an *ensemble*, as shown in Fig. 2.41. These functions may comprise, for example, records of noise, pressure fluctuations or vibration levels, taken under the same conditions but at different times.

Any quantity that cannot be precisely predicted is non-deterministic and is known as a *random variable* or a *probabilistic quantity*; that is, if a series of tests is conducted to find the value of a particular parameter x and that value is found to vary in an unpredictable way that is not a function of any other parameter, then x is a random variable.

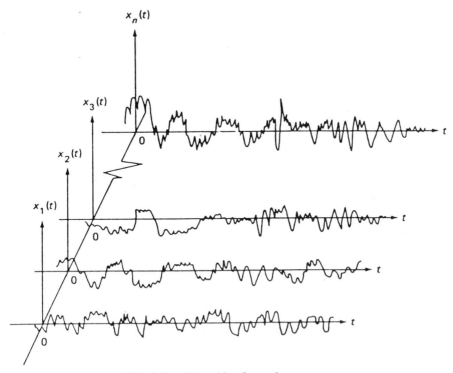

Fig. 2.41.   Ensemble of a random process.

2.3.9.1 Probability distribution

If n experimental values of a variable x are x_1, x_2, x_3, ..., x_n, the probability that the value of x will be less than x' is n'/n, where n' is the number of x values that are less than or equal to x'; that is,

$$\text{Prob } (x \leqslant x') = n'/n.$$

When n approaches ∞ this expression is the probability distribution function of x, denoted by $P(x)$, so that

$$P(x) = \underset{n \to \infty}{\text{Lt}} (n'/n)$$

The typical variation of $P(x)$ with x is shown in Fig. 2.42. Since $x(t)$ denotes a physical quantity,

$$\text{Prob } (x < -\infty) = 0, \text{ and Prob } (x < +\infty) = 1.$$

The *probability density function* is the derivative of $P(x)$ with respect to x and this is

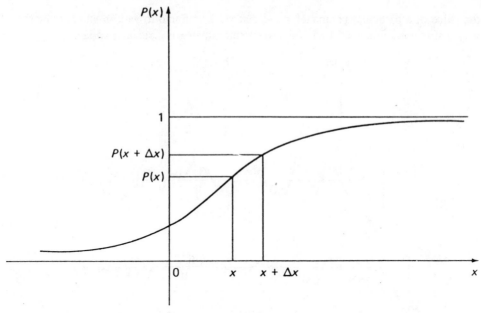

Fig. 2.42.   Probability distribution function as a function of x.

denoted by $p(x)$; that is,

$$p(x) = \frac{dP(x)}{d(x)}$$

$$= \underset{\Delta x \to 0}{\text{Lt}} \left[\frac{P(x + \Delta x) - P(x)}{\Delta x} \right],$$

where $P(x + \Delta x) - P(x)$ is the probability that the value of $x(t)$ will lie between x and $x + \Delta x$ (Fig. 2.42). Now,

$$p(x) = \frac{dP(x)}{d(x)},$$

so that

$$P(x) = \int_{-\infty}^{x} p(x)dx.$$

Hence

$$P(\infty) = \int_{-\infty}^{\infty} p(x)dx = 1,$$

so that the area under the probability density function curve is unity.

A random process is *stationary* if the joint probability density

$$p(x(t_1), x(t_2), x(t_3), \ldots)$$

depends only upon the time differences $t_2 - t_1$, $t_3 - t_2$ and so on, and not on the actual time instants; that is, the ensemble will look just the same if the time origin is changed. A random process is *ergodic* if every sample function is typical of the entire group.

The expected value of $f(x)$, which is written

$$E[f(x)] \text{ or } \overline{f(x)},$$

is

$$E[f(x)] = \overline{f(x)} = \int_{-\infty}^{\infty} f(x)p(x)dx,$$

so that the expected value of a stationary random process $x(t)$ is

$$E[x(t_1)] = E[x(t_1 + t)]$$

for any value of t.

If $f(x) = x$, the expected value or *mean value* of x,

$$E[x] \text{ or } \bar{x}, \text{ is}$$

$$E[x] = \bar{x} = \int_{-\infty}^{\infty} xp(x)dx.$$

In addition, if $f(x) = x^2$, the *mean square value* $\bar{x}^2$ of x is

$$E[x^2] = \bar{x}^2 = \int_{-\infty}^{\infty} x^2 p(x)dx.$$

The *variance* of x, σ^2, is the mean square value of x about the mean, that is,

$$\sigma^2 = E[(x - \bar{x})^2] = \int_{-\infty}^{\infty} (x - \bar{x})^2 p(x)dx = (\bar{x}^2) - (\bar{x})^2.$$

σ is the *standard deviation* of x, hence

$$\text{variance} = (\text{standard deviation})^2 = \{\text{mean square} - (\text{mean})^2\}$$

If two (or more) random variables x_1 and x_2, represent a random process at two different instants of time, then

$$E[f(x_1,x_2)] = \int_{-\infty}^{\infty} \int_{-\infty}^{\infty} f(x_1,x_2)p(x_1,x_2)dx_1 dx_2,$$

and if t_1 and t_2 are the two instants of time,

$$E[x(t_1),x(t_2)] = R(t_1,t_2),$$

which is the *auto-correlation function* for the random process (Fig. 2.43).

For random processes that are stationary,

$$E[x(t_1), x(t_2)] = R(t_1, t_2) = R(t_2 - t_1) = R(\tau),$$

say, since the average depends only upon time differences. If the process is also ergodic,

then

$$R(\tau) = \underset{T\to\infty}{\mathrm{Lt}} \frac{1}{2\pi}\int_{-T}^{T} x(t)x(t + \tau)\mathrm{d}t.$$

It is worth noting that

$$R(0) = E[x(t)^2] = \underset{T\to\infty}{\mathrm{Lt}} \frac{1}{2T}\int_{-T}^{T} x^2(t)\mathrm{d}t,$$

which is the average power in a sample function.

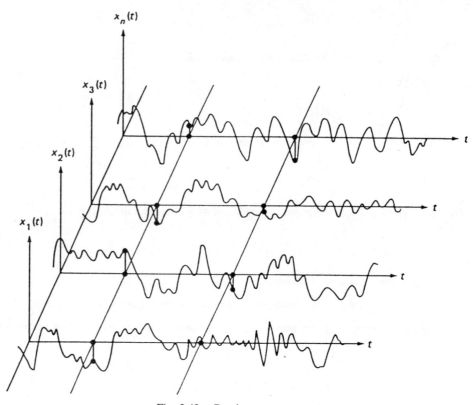

Fig. 2.43.   Random processes.

2.3.9.2 Random processes

The most important random process is the *Gaussian*, or normal random process. This is because a wide range of physically observed random waveforms can be represented as Gaussian processes, and the process has mathematical features which make analysis relatively straightforward.

The probability density function of a Gaussian process $x(t)$ is

$$p(x) = \frac{1}{\sqrt{(2\pi)}\sigma}\exp\left[-\frac{1}{2}\left(\frac{x-\bar{x}}{\sigma}\right)^2\right],$$

where σ is the standard deviation of x, and $\bar{x}$ is the mean value of x.

The values of σ and $\bar{x}$ may vary with time for a non-stationary process but are independent of time if the process is stationary.

One of the most important features of the Gaussian process is that the response of a linear system to this form of excitation is usually another, but still Gaussian, random process. The only changes are that the magnitude and standard deviation of the response may differ from those of the excitation.

A Gaussian probability density function is shown in Fig. 2.44. It can be seen to be symmetric about the mean value $\bar{x}$, and the standard deviation σ controls the spread.

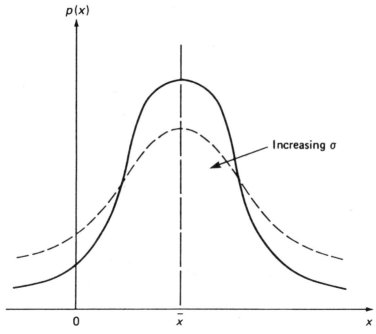

Fig. 2.44.   Gaussian probability density function.

The probability that $x(t)$ lies between $-\lambda\sigma$ and $+\lambda\sigma$, where λ is a positive number, can be found since, if $\bar{x} = 0$,

$$\text{Prob}\{-\lambda\sigma \leq x(t) \leq +\lambda\sigma\} = \int_{-\lambda\sigma}^{+\lambda\sigma}\frac{1}{\sqrt{(2\pi)}\sigma}\exp\left(-\frac{1}{2}\frac{x^2}{\sigma^2}\right)dx.$$

Fig. 2.45 shows the Gaussian probability density function with zero mean.

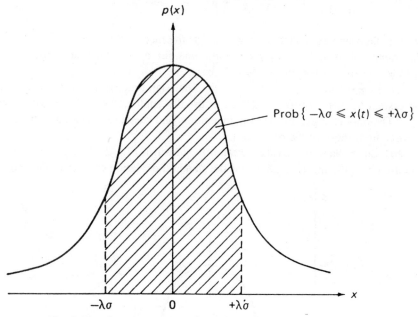

Fig. 2.45.   Gaussian probability density function with zero mean.

This integral has been calculated for a range of values of λ; the results are given in the table opposite. The probability that $x(t)$ lies outside the range $-\lambda\sigma$ to $+\lambda\sigma$ is 1 minus the value of the above integral; this probability is also given.

2.3.9.3 Spectral density

The spectral density $S(\omega)$ of a stationary random process is the Fourier transform of the autocorrelation function $R(\tau)$. It is given by

$$S(\omega) = \frac{1}{2\pi} \int_{-\infty}^{\infty} R(\tau)e^{-j\omega\tau}d\tau.$$

The inverse, which also holds true, is

$$R(\tau) = \int_{-\infty}^{\infty} S(\omega)e^{-j\omega\tau}d\omega.$$

If $\tau = 0$

$$R(0) = \int_{-\infty}^{\infty} S(\omega)d\omega = E[x^2],$$

that is, the mean square value of a stationary random process x is the area under the $S(\omega)$ against frequency curve. A typical spectral density function is shown in Fig. 2.46.

Value of λ	Prob $[-\lambda\sigma \leqslant x(t) \leqslant \lambda\sigma]$	Prob $[|x(t)| > \lambda\sigma]$
0	0	1.0000
0.2	0.1585	0.8415
0.4	0.3108	0.6892
0.6	0.4515	0.5485
0.8	0.5763	0.4237
1.0	0.6827	0.3173
1.2	0.7699	0.2301
1.4	0.8586	0.1414
1.6	0.8904	0.1096
1.8	0.9281	0.0719
2.0	0.9545	0.0455
2.2	0.9722	0.0278
2.4	0.9835	0.0164
2.6	0.9907	0.0093
2.8	0.9949	0.0051
3.0	0.9973	0.0027
3.2	0.9986	0.00137
3.4	0.9993	0.00067
3.6	0.9997	0.00032
3.8	0.9998	0.00014
4.0	0.9999	0.00006

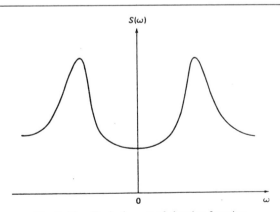

Fig. 2.46. Typical spectral density function.

A random process whose spectral density is constant over a very wide frequency range is called *white noise*. If the spectral density of a process has a significant value over a narrower range of frequencies, but one that is nevertheless still wide compared with the centre frequency of the band, it is termed a *wide-band process* (Fig. 2.47). If the frequency range is narrow compared with the centre frequency it is termed a *narrow-band process* (Fig. 2.48). Narrow-band processes frequently occur in engineering practice because real systems often respond strongly to specific exciting frequencies and thereby effectively act as a filter.

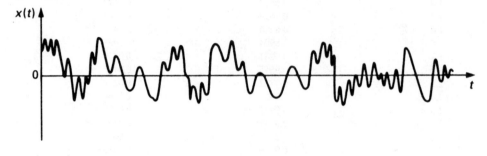

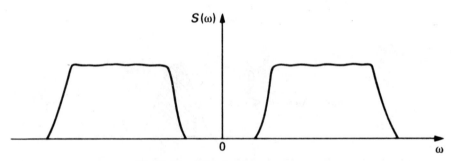

Fig. 2.47.   Wide-band process.

2.3.10 The measurement of vibration

The most commonly used device for vibration measurement is the piezoelectric accelerometer, which gives an electric signal proportional to the vibration acceleration. This signal can readily be amplified, analysed, displayed, recorded, and so on. The principles of this device can be studied by referring to Fig. 2.49 which shows a body of mass m supported by an elastic system of stiffness k and effective viscous damping of coefficient c.

This dynamic system is usually enclosed in a case which is fastened to the surface whose vibration is to be measured. The body has a pointer fixed to it, which moves over a scale fastened to the case, that is, it measures u, the motion of the suspended body relative to that of the vibrating surface so that $u = x - y$.

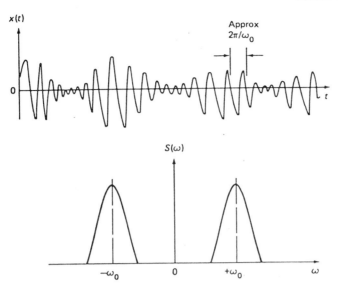

Fig. 2.48.   Narrow-band process.

Now from section 2.3.2, the amplitude of u is

$$U = \frac{A\left(\dfrac{v}{\omega}\right)^2}{\sqrt{\left\{\left[1 - \left(\dfrac{v}{\omega}\right)^2\right]^2 + \left[2\zeta\left(\dfrac{v}{\omega}\right)\right]^2\right\}}}.$$

so that if ω is low and $v \gg \omega$,

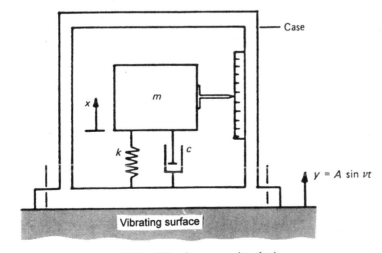

Fig. 2.49.   Vibration measuring device.

$$U \simeq \frac{A(v/\omega)^2}{(v/\omega)^2} = A,$$

that is, the device measures the input vibration amplitude; when operating in this mode it is called a *vibrometer*, and if ω is high so that $\omega \gg v$, then

$$U \simeq \frac{A(v/\omega)^2}{1} = \frac{1}{\omega^2}Av^2,$$

that is, the device measures the input vibration acceleration amplitude; when operating in this mode it is called an *accelerometer*.

By adjusting the system parameters correctly it is possible to make

$$\sqrt{\left\{\left[1 - \left(\frac{v}{\omega}\right)^2\right]^2 + \left[2\zeta\left(\frac{v}{\omega}\right)\right]^2\right\}}$$

have a value close to unity for exciting frequencies v up to about 0.3ω. Commercial accelerometers usually have piezoelectric elements instead of a spring and damper, so that the electric signal produced is proportional to the relative motion, u, above.

Piezoelectric accelerometers are widely used for measuring the vibration of structures. The output of these accelerometers is governed by their sensitivity; in general the larger and therefore the heavier the accelerometer, the greater its sensitivity and the greater the output for a given excitation g-level. However, accelerometers have to be attached to the structure and large accelerometers may affect the response of the structure due to their added mass and they may also have a limited frequency range. Smaller accelerometers have stiffer piezoelectric elements which are less sensitive but can operate at higher frequencies. The output of piezoelectric accelerometers is easily amplified, analysed and recorded.

Strain gauges are also often used to measure the dynamic response of a structure. These rely on the change in resistance of a wire caused by a change in its length. Dynamic measurements require using an a.c. bridge circuit, the carrier frequency used determining the range of frequency measurements possible. These gauges are cheap and easy to apply to a structure, and the bridge output is easily recorded and analysed.

Non-contacting capacitance and impedance transducers are also sometimes used.

3

The vibration of structures with more than one degree of freedom

Many real structures can be represented by a single degree of freedom model. However, most actual structures have several bodies and several restraints and therefore several degrees of freedom. The number of degrees of freedom that a structure possesses is equal to the number of independent coordinates necessary to describe the motion of the system. Since no body is completely rigid, and no spring is without mass, every real structure has more than one degree of freedom, and sometimes it is not sufficiently realistic to approximate a structure by a single degree of freedom model. Thus, it is necessary to study the vibration of structures with more than one degree of freedom.

Each flexibly connected body in a multi-degree of freedom structure can move independently of the other bodies, and only under certain conditions will all bodies undergo a harmonic motion at the same frequency. Since all bodies move with the same frequency, they all attain their amplitudes at the same time even if they do not all move in the same direction. When such motion occurs the frequency is called a *natural frequency* of the structure and the motion is a principal mode of vibration: the number of natural frequencies and principal modes that a structure possesses is equal to the number of degrees of freedom of that structure. The deployment of the structure at its lowest or first natural frequency is called its first mode, at the next highest or second natural frequency it is called the second mode, and so on.

A two degree of freedom structure will be considered initially. This is because the addition of more degrees of freedom increases the labour of the solution procedure but does not introduce any new analytical principles.

Initially, we will obtain the equations of motion for a two degree of freedom model, and from these find the natural frequencies and corresponding mode shapes.

Some examples of two degree of freedom models of vibrating structures are shown in Figs 3.1(a)–(e).

3.1   THE VIBRATION OF STRUCTURES WITH TWO DEGREES OF FREEDOM

3.1.1 Free vibration of an undamped structure

Of the examples of two degree of freedom models shown in Fig. 3.1(a)–(e), consider the system shown in Fig. 3.1(a). If $x_1 > x_2$ the FBDs are as shown in Fig. 3.2.

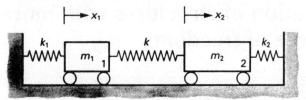

Fig. 3.1(a).   Linear undamped system, horizontal motion. Coordinates x_1 and x_2.

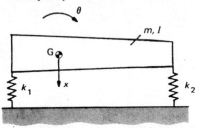

Fig. 3.1(b).   System with combined translation and rotation. Coordinates x and θ.

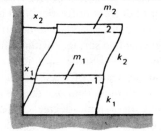

Fig. 3.1(c).   Shear frame. Coordinates x_1 and x_2.

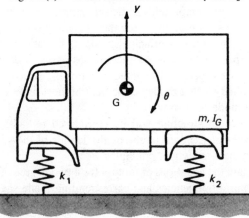

Fig. 3.1(d).   Two degree of freedom model, rotation plus translation. Coordinates y and θ.

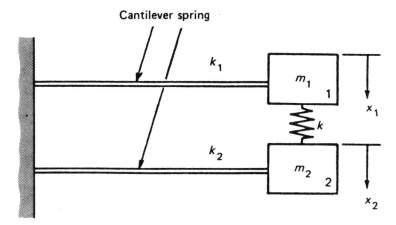

Fig. 3.1(e).   Two degree of freedom model, translation vibration. Coordinates x_1 and x_2.

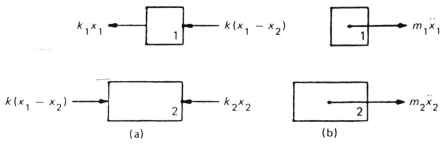

(a)   (b)

Fig. 3.2.   (a) Applied forces, (b) effective forces.

The equations of motion are therefore,

$$m_1\ddot{x}_1 = -k_1x_1 - k(x_1 - x_2) \quad \text{for body 1,} \tag{3.1}$$

and

$$m_2\ddot{x}_2 = k(x_1 - x_2) - k_2x_2 \quad \text{for body 2.} \tag{3.2}$$

The same equations are obtained if $x_1 < x_2$ is assumed because the direction of the central spring force is then reversed.

Equations (3.1) and (3.2) can be solved for the natural frequencies and corresponding mode shapes by assuming a solution of the form

$$x_1 = A_1\sin(\omega t + \psi) \text{ and } x_2 = A_2\sin(\omega t + \psi).$$

This assumes that x_1 and x_2 oscillate with the same frequency ω and are either in phase or π out of phase. This is a sufficient condition to make ω a natural frequency.

Substituting these solutions into the equations of motion gives

$$-m_1 A_1 \omega^2 \sin(\omega t + \psi) = -k_1 A_1 \sin(\omega t + \psi) - k(A_1 - A_2)\sin(\omega t + \psi)$$

and

$$-m_2 A_2 \omega^2 \sin(\omega t + \psi) = k(A_1 - A_2)\sin(\omega t + \psi) - k_2 A_2 \sin(\omega t + \psi).$$

Since these solutions are true for all values of t,

$$A_1(k + k_1 - m_1 \omega^2) + A_2(-k) = 0 \tag{3.3}$$

and

$$A_1(-k) + A_2(k_2 + k - m_2 \omega^2) = 0. \tag{3.4}$$

A_1 and A_2 can be eliminated by writing

$$\begin{vmatrix} k + k_1 - m_1 \omega^2 & -k \\ -k & k + k_2 - m_2 \omega^2 \end{vmatrix} = 0 \tag{3.5}$$

This is the *characteristic* or *frequency equation*. Alternatively, we may write

$$A_1/A_2 = k/(k + k_1 - m_1 \omega^2) \text{ from } (3.3)$$

and

$$A_1/A_2 = (k_2 + k - m_2 \omega^2)/k \text{ from } (3.4) \tag{3.6}$$

Thus

$$k/(k + k_1 - m_1 \omega^2) = (k_2 + k - m_2 \omega^2)/k$$

and

$$(k + k_1 - m_1 \omega^2)(k_2 + k - m_2 \omega^2) - k^2 = 0. \tag{3.7}$$

This result is the frequency equation and could also be obtained by multiplying out the above determinant (equation (3.5)).

The solutions to equation (3.7) give the natural frequencies of free vibration for the system in Fig. 3.1(a). The corresponding mode shapes are found by substituting these frequencies, in turn, into either of equations (3.6).

Consider the case when $k_1 = k_2 = k$, and $m_1 = m_2 = m$. The frequency equation is $(2k - m\omega^2)^2 - k^2 = 0$; that is,

$$m^2 \omega^4 - 4mk\omega^2 + 3k^2 = 0, \text{ or } (m\omega^2 - k)(m\omega^2 - 3k) = 0.$$

Therefore, either $m\omega^2 - k = 0$ or $m\omega^2 - 3k = 0$.

Thus

$$\omega_1 = \sqrt{(k/m)} \text{ rad/s} \quad \text{and} \quad \omega_2 = \sqrt{(3k/m)} \text{ rad/s}.$$

If

$$\omega = \sqrt{(k/m)} \text{ rad/s}, \quad (A_1/A_2)_{\omega = \sqrt{(k/m)}} = + 1,$$

and if

$$\omega = \sqrt{(3k/m)} \text{ rad/s}, \quad (A_1/A_2)_{\omega = \sqrt{(3k/m)}} = -1. \qquad \text{(from 3.6)}$$

This gives the mode shapes corresponding to the frequencies ω_1 and ω_2. Thus, the first mode of free vibration occurs at a frequency $(1/2\pi)\sqrt{(k/m)}$ Hz and $(A_1/A_2)^{\mathrm{I}} = 1$, that is, the bodies move in phase with each other and with the same amplitude as if connected by a rigid link (Fig. 3.3). The second mode of free vibration occurs at a frequency $(1/2\pi)\sqrt{(3k/m)}$ Hz and $(A_1/A_2)^{\mathrm{II}} = -1$, that is, the bodies move exactly out of phase with each other, but with the same amplitude (see Fig. 3.3).

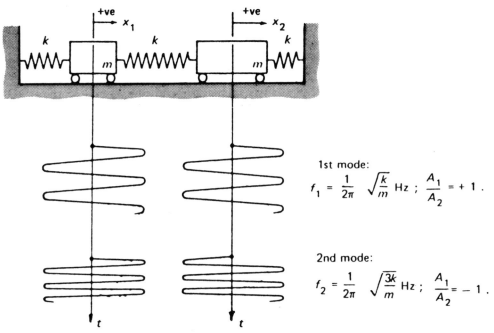

1st mode:

$$f_1 = \frac{1}{2\pi} \sqrt{\frac{k}{m}} \text{ Hz} ; \quad \frac{A_1}{A_2} = + 1 .$$

2nd mode:

$$f_2 = \frac{1}{2\pi} \sqrt{\frac{3k}{m}} \text{ Hz} ; \quad \frac{A_1}{A_2} = - 1 .$$

Fig. 3.3.   Natural frequencies and mode shapes for two degree of freedom translation vibration system. Bodies of equal mass and springs of equal stiffness.

3.1.1.1 Free motion

The two modes of vibration can be written

$$\begin{Bmatrix} x_1 \\ x_2 \end{Bmatrix}^{\mathrm{I}} = \begin{Bmatrix} A_1 \\ A_2 \end{Bmatrix}^{\mathrm{I}} \sin(\omega_1 t + \psi_1)$$

and

$$\left\{ \frac{x_1}{x_2} \right\}^{II} = \left\{ \frac{A_1}{A_2} \right\}^{II} \sin(\omega_2 t + \psi_2),$$

where the ratio A_1/A_2 is specified for each mode.

Since each solution satisfies the equation of motion, the general solution is

$$\left\{ \frac{x_1}{x_2} \right\} = \left\{ \frac{A_1}{A_2} \right\}^{I} \sin(\omega_1 t + \psi_1) + \left\{ \frac{A_1}{A_2} \right\}^{II} \sin(\omega_2 t + \psi_2),$$

where A_1, A_2, ψ_1, ψ_2 are found from the initial conditions.

For example, for the system considered above, if one body is displaced a distance X and released,

$$x_1(0) = X \text{ and } x_2(0) = \dot{x}_1(0) = \dot{x}_2(0) = 0,$$

where $x_1(0)$ means the value of x_1 when $t = 0$, and similarly for $x_2(0)$, $\dot{x}_1(0)$ and $\dot{x}_2(0)$. Remembering that in this system $\omega_1 = \sqrt{(k/m)}$, $\omega_2 = \sqrt{(3k/m)}$, and

$$\left(\frac{A_1}{A_2} \right)_{\omega_1} = +1 \quad \text{and} \quad \left(\frac{A_1}{A_2} \right)_{\omega_2} = -1,$$

we can write

$$x_1 = \sin(\sqrt{(k/m)}t + \psi_1) + \sin(\sqrt{(3k/m)}t + \psi_2),$$

and

$$x_2 = \sin(\sqrt{(k/m)}t + \psi_1) - \sin(\sqrt{(3k/m)}t + \psi_2).$$

Substituting the initial conditions $x_1(0) = X$ and $x_2(0) = 0$ gives

$$X = \sin \psi_1 + \sin \psi_2$$

and

$$0 = \sin \psi_1 - \sin \psi_2,$$

that is,

$$\sin \psi_1 = \sin \psi_2 = X/2.$$

The remaining conditions give $\cos \psi_1 = \cos \psi_2 = 0$.
Hence

$$x_1 = (X/2) \cos \sqrt{(k/m)}t + (X/2)\cos\sqrt{(3k/m)}t,$$

and

$$x_2 = (X/2)\cos\sqrt{(k/m)}t - (X/2)\cos\sqrt{(3k/m)}t.$$

That is, *both* natural frequencies are excited and the motion of each body has two harmonic components.

3.1.2 Coordinate coupling

In some structures the motion is such that the coordinates are coupled in the equations of motion. Consider the system shown in Fig. 3.1(b); only motion in the plane of the figure is considered, horizontal motion being neglected because the lateral stiffness of the springs is assumed to be negligible. The coordinates of rotation, θ, and translation, x, are coupled as shown in Fig. 3.4. G is the centre of mass of the rigid beam of mass m and moment of inertia I about G.

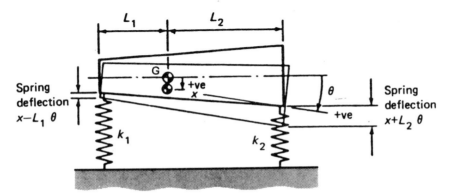

Fig. 3.4.   Two degree of freedom model, rotation plus translation.

The FBDs are shown in Fig. 3.5; since the weight of the beam is supported by the springs, both the initial spring forces and the beam weight may be omitted.

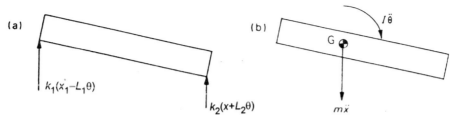

Fig. 3.5.   (a) Applied forces, (b) effective force and moment.

For small amplitudes of oscillation (so that $\sin\theta \simeq \theta$) the equations of motion are

$$m\ddot{x} = -k_1(x - L_1\theta) - k_2(x + L_2\theta)$$

and

$$I\ddot{\theta} = k_1(x - L_1\theta)L_1 - k_2(x + L_2\theta)L_2,$$

that is,

$$m\ddot{x} + (k_1 + k_2)x - (k_1L_1 - k_2L_2)\theta = 0$$

and

$$I\ddot{\theta} - (k_1L_1 - k_2L_2)x + (k_1L_1^2 + k_2L_2^2)\theta = 0.$$

It will be noticed that these equations can be uncoupled by making $k_1L_1 = k_2L_2$; if this is arranged, translation (x motion) and rotation (θ motion) can take place independently. Otherwise translation and rotation occur simultaneously.

Assuming $x = A_1 \sin(\omega t + \psi)$ and $\theta = A_2 \sin(\omega t + \psi)$, substituting into the equations of motion gives

$$-m\omega^2 A_1 + (k_1 + k_2)A_1 - (k_1L_1 - k_2L_2)A_2 = 0$$

and

$$-I\omega^2 A_2 - (k_1L_1 - k_2L_2)A_1 + (k_1L_1^2 + k_2L_2^2)A_2 = 0,$$

that is,

$$A_1(k_1 + k_2 - m\omega^2) + A_2(-(k_1L_1 - k_2L_2)) = 0$$

and

$$A_1(-(k_1L_1 - k_2L_2)) + A_2(k_1L_1^2 + k_2L_2^2 - I\omega^2) = 0.$$

Hence the frequency equation is

$$\begin{vmatrix} k_1 + k_2 - m\omega^2 & -(k_1L_1 - k_2L_2) \\ -(k_1L_1 - k_2L_2) & k_1L_1^2 + k_2L_2^2 - I\omega^2 \end{vmatrix} = 0.$$

For each natural frequency, there is a corresponding mode shape, given by A_1/A_2.

Example 18

When transported, a space vehicle is supported in a horizontal position by two springs, as shown. The vehicle can be considered to be a rigid body of mass m and radius of gyration h about an axis normal to the plane of the figure through the mass centre G. The rear support has a stiffness k_1 and is at a distance a from G while the front support has a stiffness k_2 and is at a distance b from G. The only motions possible for the vehicle are vertical translation and rotation in the vertical plane.

Write the equations of small amplitude motion of the vehicle and obtain the frequency equation in terms of the given parameters.

Given that $k_1a = k_2b$, determine the natural frequencies of the free vibrations of the vehicle and sketch the corresponding modes of vibration. Also state or sketch the modes of vibration if $k_1a \neq k_2b$.

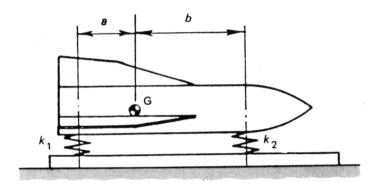

The FBDs are as below:

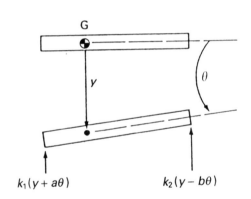

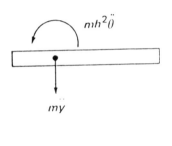

The equations of motion are

$$k_1(y + a\theta) + k_2(y - b\theta) = -m\ddot{y}$$

and

$$k_1(y + a\theta)a - k_2(y - b\theta)b = -mh^2\ddot{\theta}.$$

Assuming

$$y = Y \sin vt \quad \text{and} \quad \theta = \Theta \sin vt,$$

these give

$$Y(k_1 + k_2 - mv^2) + \Theta(k_1a - k_2b) = 0,$$

and

$$Y(k_1a - k_2b) + \Theta(k_1a^2 + k_2b^2 - mh^2v^2) = 0.$$

The frequency equation is, therefore,

$$(k_1 + k_2 - mv^2)(k_1a^2 + k_2b^2 - mh^2v^2) - (k_1a - k_2b)^2 = 0.$$

If $k_1a = k_2b$, motion is uncoupled so

$$v_1 = \sqrt{\left(\frac{k_1 + k_2}{m}\right)} \text{ rad/s} \quad \text{and} \quad v_2 = \sqrt{\left(\frac{k_1a^2 + k_2b^2}{mh^2}\right)} \text{ rad/s;}$$

v_1 is the frequency of a bouncing or translation mode (no rotation):

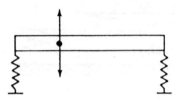

v_2 is the frequency of a rotation mode (no bounce):

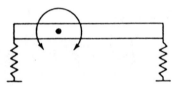

If $k_1a \neq k_2b$, the modes are coupled:

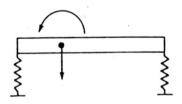 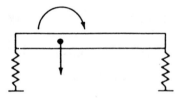

Example 19

In a study of earthquakes, a building is idealized as a rigid body of mass M supported on two springs, one giving translational stiffness k and the other rotational stiffness k_T as shown.

Given that I_G is the mass moment of inertia of the building about its mass centre G, write down the equations of motion using coordinates x for the translation from the equilibrium position, and θ for the rotation of the building.

Hence determine the frequency equation of the motion.

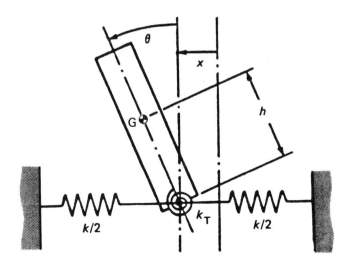

The FBDs are as follows.

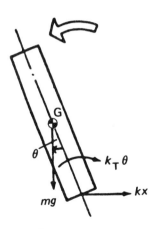
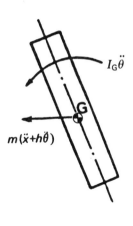

Assume small θ (earthquakes), hence

$$m(\ddot{x} + h\ddot{\theta}) = -kx,$$

and

$$I_G\ddot{\theta} + m(\ddot{x} + h\ddot{\theta})h = -k_T\theta + mgh\theta.$$

The equations of motion are therefore

$$mh\ddot{\theta} + m\ddot{x} + kx = 0,$$

and

$$mh\ddot{x} + (mh^2 + I_G)\ddot{\theta} - (mgh - k_T)\theta = 0.$$

If $\theta = A_1 \sin \omega t$ and $x = A_2 \sin \omega t$,

$$-mh\omega^2 A_1 - m\omega^2 A_2 + kA_2 = 0$$

and

$$+ mh\omega^2 A_2 + (mh^2 + I_G)\omega^2 A_1 + (mgh - k_T)A_1 = 0.$$

The frequency equation is

$$\begin{vmatrix} -mh\omega^2 & k - m\omega^2 \\ (mh^2 + I_G)\omega^2 + (mgh - k_T) & mh\omega^2 \end{vmatrix} = 0,$$

that is,

$$(mh\omega^2)^2 + (k - m\omega^2)[(mh^2 + I_G)\omega^2 + (mgh - k_T)] = 0$$

or

$$mI_G\omega^4 - \omega^2[mkh^2 + I_G k - m^2 gh + mk_T] - mghk + kk_T = 0.$$

3.1.3 Forced vibration

Harmonic excitation of vibration in a structure may be generated in a number of ways, for example by unbalanced rotating or reciprocating machinery, or it may arise from periodic excitation containing a troublesome harmonic component.

A two degree of freedom model of a structure excited by a harmonic force $F \sin vt$ is shown in Fig. 3.6. Damping is assumed to be negligible. The force has a constant amplitude F and a frequency $v/2\pi$ Hz.

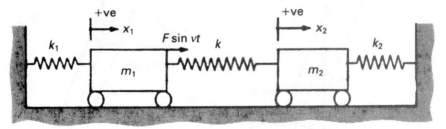

Fig. 3.6.   Two degree of freedom model with forced excitation.

The equations of motion are

$$m_1\ddot{x}_1 = - k_1 x_1 - k(x_1 - x_2) + F \sin vt,$$

and

$$m_2\ddot{x}_2 = k(x_1 - x_2) - k_2 x_2.$$

Since there is zero damping, the motions are either in phase or π out of phase with the driving force, so that the following solutions may be assumed:

$$x_1 = A_1 \sin vt \quad \text{and} \quad x_2 = A_2 \sin vt.$$

Substituting these solutions into the equations of motion gives

$$A_1(k_1 + k - m_1 v^2) + A_2(-k) = F$$

and

$$A_1(-k) + A_2(k_2 + k - m_2v^2) = 0.$$

Thus

$$A_1 = \frac{F(k_2 + k - m_2v^2)}{\Delta},$$

and

$$A_2 = \frac{Fk}{\Delta},$$

where

$$\Delta = (k_2 + k - m_2v^2)(k_1 + k - m_1v^2) - k^2$$

and $\Delta = 0$ is the frequency equation.

Hence the response of the system to the exciting force is determined.

Example 20

A two-wheel trailer is drawn over an undulating surface in such a way that the vertical motion of the tyre may be regarded as sinusoidal, the pitch of the undulations being 5 m. The combined stiffness of the tyres is 170 kN/m and that of the main springs is 60 kN/m; the axle and attached parts have a mass of 400 kg, and the mass of the body is 500 kg. Find (a) the critical speeds of the trailer in km/h and (b) the amplitude of the trailer body vibration if the trailer is drawn at 50 km/h and the amplitude of the undulations is 0.1 m.

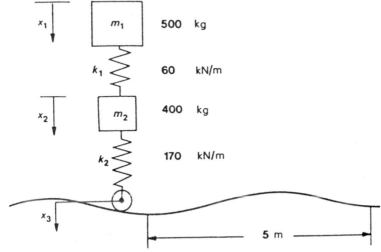

The equations of motion are

$$m_1\ddot{x}_1 = -k_1(x_1 - x_2),$$

and

$$m_2\ddot{x}_2 = k_1(x_1 - x_2) - k_2(x_2 - x_3).$$

Assuming $x_1 = A_1 \sin vt$, $x_2 = A_2 \sin vt$, and $x_3 = A_3 \sin vt$,

$$A_1(k_1 - m_1 v^2) + A_2(-k_1) = 0$$

and

$$A_1(-k_1) + A_2(k_1 + k_2 - m_2 v^2) = k_2 A_3.$$

The frequency equation is

$$(k_1 + k_2 - m_2 v^2)(k_1 - m_1 v^2) - k_1^2 = 0.$$

The critical speeds are those which correspond to the natural frequencies and hence excite resonances. The frequency equation simplifies to

$$m_1 m_2 v^4 - (m_1 k_1 + m_1 k_2 + m_2 k_1)v^2 + k_1 k_2 = 0.$$

Hence substituting the given data,

$$500 \times 400 \times v^4 - (500 \times 60 + 500 \times 170 + 400 \times 60)\ 10^3 v^2 + 60 \times 170 \times 10^6 = 0,$$

that is $0.2v^4 - 139v^2 + 10\ 200 = 0$, which can be solved by the formula. Thus $v = 16.3$ rad/s or 20.78 rad/s, and $f = 2.59$ Hz or 3.3 Hz.

Now if the trailer is drawn at v km/h, or $v/3.6$ m/s, the frequency is $v/(3.6 \times 5)$ Hz. Therefore the critical speeds are

$$v_1 = 18 \times 2.59 = 46.6 \text{ km/h},$$

and

$$v_2 = 18 \times 3.3 = 59.4 \text{ km/h}.$$

Towing the trailer at either of these speeds will excite a resonance in the system. From the equations of motion,

$$A_1 = \left\{ \frac{k_1 k_2}{(k_1 + k_2 - m_2 v^2)(k_1 - m_1 v^2) - k_1^2} \right\} A_3,$$

$$= \left\{ \frac{10\ 200}{0.2v^4 - 139v^2 + 10\ 200} \right\} A_3.$$

At 50 km/h, $v = 17.49$ rad/s.

Thus $A_1 = -0.749A_3$. Since $A_3 = 0.1$ m, the amplitude of the trailer vibration is 0.075 m. This motion is π out of phase with the road undulations.

3.1.4 Structure with viscous damping

If a structure possesses damping of a viscous nature, the damping can be modelled similarly to that in the system shown in Fig. 3.7.

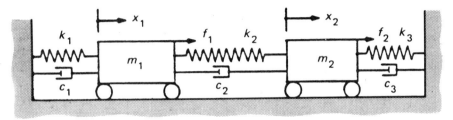

Fig. 3.7.   Two degree of freedom viscous damped model with forced excitation.

For this system the equations of motion are

$$m_1\ddot{x}_1 + k_1 x_1 + k_2(x_1 - x_2) + c_1\dot{x}_1 + c_2(\dot{x}_1 - \dot{x}_2) = f_1$$

and

$$m_2\ddot{x}_2 + k_2(x_2 - x_1) + k_3 x_2 + c_2(\dot{x}_2 - \dot{x}_1) + c_3\dot{x}_2 = f_2.$$

Solutions of the form $x_1 = A_1 e^{st}$ and $x_2 = A_2 e^{st}$ can be assumed, where the Laplace operator is equal to $a + jb$, $j = \sqrt{(-1)}$, and a and b are real — that is, each solution contains a harmonic component of frequency b, and a vibration decay component of damping factor a. By substituting these solutions into the equations of motion a frequency equation of the form

$$s^4 + \alpha s^3 + \beta s^2 + \gamma s + \delta = 0$$

can be deduced, where α, β, γ and δ are real coefficients. From this equation four roots and thus four values of s can be obtained. In general the roots form two complex conjugate pairs such as $a_1 \pm jb_1$, and $a_2 \pm jb_2$. These represent solutions of the form $x = \text{Re}(Xe^{at} . e^{jbt}) = Xe^{at}\cos bt$; that is the motion of the bodies is harmonic, and decays exponentially with time. The parameters of the system determine the magnitude of the frequency and the decay rate.

It is often convenient to plot these roots on a complex plane as shown in Fig. 3.8. This is known as the s-plane.

For light damping the damped frequency for each mode is approximately equal to the undamped frequency, that is, $b_1 \simeq \omega_1$ and $b_2 \simeq \omega_2$.

The right-hand side of the s-plane (Re(s) +ve) represents a root with a positive exponent, that is, a term that grows with time, so unstable motion may exist. The left-hand side contains roots with a negative exponent so stable motion exists.

All passive systems have negative real parts and are therefore stable but some systems such as rolling wheels and rockets can become unstable, and thus it is important that the stability of a system is considered. This can be conveniently done by plotting the roots of the frequency equation on the s-plane.

3.1.5 Structures with other forms of damping

For most structures the level of damping is such that the damped natural frequencies are very nearly equal to the undamped natural frequencies. Thus, if only the natural frequencies of the structure are required, damping can usually be neglected in the analysis. This is a significant simplification. Also, if the response of a structure at a frequency well

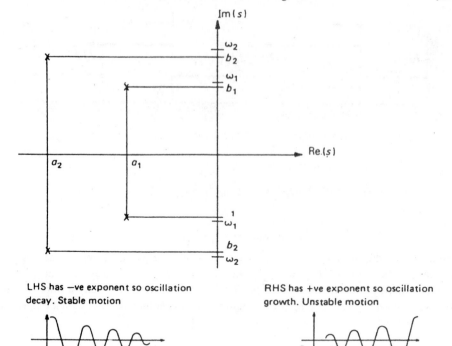

Fig. 3.8.   *s*-plane.

away from a resonance is required, a similar simplification may be made in the analysis.

However, if the response of a structure at a frequency in the region of a resonance is required, which would be the case if the amplitude or dynamic stress levels at a resonance were required for example, damping effects must be included in the analysis.

Coulomb and hysteretic damping can be difficult to analyse exactly, particularly in multi-degree of freedom systems, but approximations can be made to linearize the equations of motion. For example, an equivalent viscous damping coefficient for equal energy dissipation may be assumed as shown in section 2.2.6, or alternatively the non-linear damping force may be replaced by an equivalent harmonic force or series of forces, as discussed in section 2.3.8.

3.2   THE VIBRATION OF STRUCTURES WITH MORE THAN TWO DEGREES OF FREEDOM

The vibration analysis of a structure with three or more degrees of freedom can be carried out in the same way as the analysis given above for two degrees of freedom. However, the method becomes tedious for many degrees of freedom, and numerical methods may have to be used to solve the frequency equation. A computer can, of course, be used to solve the frequency equation and determine the corresponding mode shapes. Although computational and computer techniques are extensively used in the analysis of multi-degree of freedom structures, it is essential for the analytical and numerical bases of any program used to be understood, to ensure its relevance to the problem considered, and that the program does not introduce unacceptable approximations and calculation errors. For this reason it is necessary to derive the basic theory and equations for multi-degree of freedom structures. Computational techniques are essential, and widely used, for the analysis of the sophisticated structural models often devised and considered necessary, and computer packages are available for routine analyses. However, considerable economies in writing the analysis and performing the computations can be achieved, by adopting a matrix method for the analysis. Alternatively an energy solution can be obtained by using the Lagrange equation, or some simplification in the analysis achieved by using the receptance technique. The matrix method will be considered first.

3.2.1 The matrix method

The matrix method for analysis is a convenient way of handling several equations of motion. Furthermore, specific information about a structure such as its lowest natural frequency, can be obtained without carrying out a complete and detailed analysis. The matrix method of analysis is particularly important because it forms the basis of many computer solutions to vibration problems. The method can best be demonstrated by means of an example. For a full description of the matrix method see *Mechanical Vibrations: Introduction to Matrix Methods* by J. M. Prentis & F. A. Leckie (Longmans, 1963).

Example 21

A structure is modelled by the three degree of freedom system shown. Determine the highest natural frequency of free vibration and the associated mode shape.

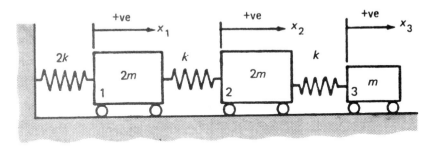

The equations of motion are

$$2m\ddot{x}_1 + 2kx_1 + k(x_1 - x_2) = 0,$$

$$2m\ddot{x}_2 + k(x_2 - x_1) + k(x_2 - x_3) = 0$$

and

$$m\ddot{x}_3 + k(x_3 - x_2) = 0.$$

If x_1, x_2 and x_3 take the form $X \sin \omega t$ and $\lambda = m\omega^2/k$, these equations can be written

$$\tfrac{3}{2}X_1 - \tfrac{1}{2}X_2 \qquad = \lambda X_1,$$

$$-\tfrac{1}{2}X_1 + X_2 - \tfrac{1}{2}X_3 = \lambda X_2$$

and

$$-X_2 + X_3 \qquad = \lambda X_3,$$

that is,

$$\begin{bmatrix} 1.5 & -0.5 & 0 \\ -0.5 & 1 & -0.5 \\ 0 & -1 & 1 \end{bmatrix} \begin{Bmatrix} X_1 \\ X_2 \\ X_3 \end{Bmatrix} = \lambda \begin{Bmatrix} X_1 \\ X_2 \\ X_3 \end{Bmatrix}$$

or

$$[S]\{X\} = \lambda\{X\}$$

where $[S]$ is the system matrix, $\{X\}$ is a column matrix, and the factor λ is a scalar quantity.

This matrix equation can be solved by an iteration procedure. This procedure is started by assuming a set of deflections for the column matrix $\{X\}$ and multiplying by $[S]$; this results in a new column matrix. This matrix is normalized by making one of the amplitudes unity and dividing each term in the column by the particular amplitude which was put equal to unity. The procedure is repeated until the amplitudes stabilize to a definite pattern. Convergence is always to the highest value of λ and its associated column matrix. Since $\lambda = m\omega^2/k$, this means that the highest natural frequency is found. Thus to start the iteration a reasonable assumed mode would be

$$\begin{Bmatrix} X_1 \\ X_2 \\ X_3 \end{Bmatrix} = \begin{Bmatrix} 1 \\ -1 \\ 2 \end{Bmatrix}.$$

Now

$$\begin{bmatrix} 1.5 & -0.5 & 0 \\ -0.5 & 1 & -0.5 \\ 0 & -1 & 1 \end{bmatrix} \begin{Bmatrix} 1 \\ -1 \\ 2 \end{Bmatrix} = \begin{Bmatrix} 2 \\ -2.5 \\ 3 \end{Bmatrix} = 3 \begin{Bmatrix} 0.67 \\ -0.83 \\ 1.00 \end{Bmatrix}$$

Using this new column matrix gives

$$\begin{bmatrix} 1.5 & -0.5 & 0 \\ -0.5 & 1 & -0.5 \\ 0 & -1 & 1 \end{bmatrix} \begin{Bmatrix} 0.67 \\ -0.83 \\ 1.00 \end{Bmatrix} = \begin{Bmatrix} 1.415 \\ -1.665 \\ 1.83 \end{Bmatrix} = 1.83 \begin{Bmatrix} 0.77 \\ -0.91 \\ 1.00 \end{Bmatrix}$$

and eventually, by repeating the process the following is obtained:

$$\begin{bmatrix} 1.5 & -0.5 & 0 \\ -0.5 & 1 & -0.5 \\ 0 & -1 & 1 \end{bmatrix} \begin{Bmatrix} 1 \\ -1 \\ 1 \end{Bmatrix} = 2 \begin{Bmatrix} 1 \\ -1 \\ 1 \end{Bmatrix}$$

Hence $\lambda = 2$ and $\omega^2 = 2k/m$. λ is an *eigenvalue* of $[S]$, and the associated value of $\{X\}$ is the corresponding *eigenvector* of $[S]$. The eigenvector gives the mode shape.

Thus the highest natural frequency is $1/2\pi \sqrt{(2k/m)}$ Hz, and the associated mode shape is $1: -1:1$. Thus if $X_1 = 1$, $X_2 = -1$ and $X_3 = 1$.

If the lowest natural frequency is required, it can be found from the lowest eigenvalue. This can be obtained directly by inverting [S] and premultiplying $[S]\{X\} = \lambda\{X\}$ by $\lambda^{-1}[S]^{-1}$.

Thus $[S]^{-1}\{X\} = \lambda^{-1}\{X\}$. Iteration of this equation yields the largest value of λ^{-1} and hence the lowest natural frequency. A reasonable assumed mode for the first iteration would be

$$\begin{Bmatrix} 1 \\ 1 \\ 2 \end{Bmatrix}.$$

Alternatively, the lowest eigenvalue can be found from the flexibility matrix. The flexibility matrix is written in terms of the influence coefficients. The influence coefficient α_{pq} of a system is the deflection (or rotation) at the point p due to a unit force (or moment) applied at a point q. Thus, since the force each body applies is the product of its mass and acceleration:

$$X_1 = \alpha_{11}\, 2mX_1\omega^2 + \alpha_{12}\, 2mX_2\omega^2 + \alpha_{13}mX_3\omega^2,$$

$$X_2 = \alpha_{21}\, 2mX_1\omega^2 + \alpha_{22}\, 2mX_2\omega^2 + \alpha_{23}mX_3\omega^2,$$

and

$$X_3 = \alpha_{31}\, 2mX_1\omega^2 + \alpha_{32}\, 2mX_2\omega^2 + \alpha_{33}mX_3\omega^2,$$

or

$$\begin{bmatrix} 2\alpha_{11} & 2\alpha_{12} & \alpha_{13} \\ 2\alpha_{21} & 2\alpha_{22} & \alpha_{23} \\ 2\alpha_{31} & 2\alpha_{32} & \alpha_{33} \end{bmatrix} \begin{Bmatrix} X_1 \\ X_2 \\ X_3 \end{Bmatrix} = \frac{1}{m\omega^2} \begin{Bmatrix} X_1 \\ X_2 \\ X_3 \end{Bmatrix}.$$

The influence coefficients are calculated by applying a unit force or moment to each body in turn. Since the same unit force acts between the support and its point of application, the displacement of the point of application of the force is the sum of the extensions of the springs extended. The displacements of all points beyond the point of application of the force are the same.

Thus

$$\alpha_{11} = \alpha_{12} = \alpha_{13} = \alpha_{21} = \alpha_{31} = \frac{1}{2k},$$

$$\alpha_{22} = \alpha_{23} = \alpha_{32} = \frac{1}{2k} + \frac{1}{k} = \frac{3}{2k},$$

and

$$\alpha_{33} = \frac{1}{2k} + \frac{1}{k} + \frac{1}{k} = \frac{5}{2k}.$$

Iteration causes the eigenvalue $k/m\omega^2$ to converge to its highest value, and hence the lowest natural frequency is found. The other natural frequencies of the system can be found by applying the orthogonality relation between the principal modes of vibration.

3.2.1.1 Orthogonality of the principal modes of vibration

Consider a linear elastic system that has n degrees of freedom, n natural frequencies and n principal modes.

The orthogonality relation between the principal modes of vibration for an n degree of freedom system is

$$\sum_{i=1}^{n} m_i A_i(r)A_i(s) = 0,$$

where $A_i(r)$ are the amplitudes corresponding to the rth mode, and $A_i(s)$ are the amplitudes corresponding to the sth mode.

This relationship is used to sweep unwanted modes from the system matrix, as illustrated in the following example.

Example 22

Consider the three degree of freedom model of a structure shown.

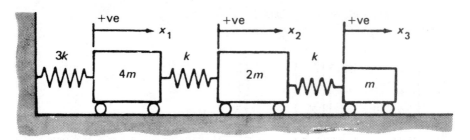

The equations of motion in terms of the influence coefficients are

$$X_1 = 4\alpha_{11}mX_1\omega^2 + 2\alpha_{12}mX_2\omega^2 + \alpha_{13}mX_3\omega^2,$$
$$X_2 = 4\alpha_{21}mX_1\omega^2 + 2\alpha_{22}mX_2\omega^2 + \alpha_{23}mX_3\omega^2$$

and

$$X_3 = 4\alpha_{31}mX_1\omega^2 + 2\alpha_{32}mX_2\omega^2 + \alpha_{33}mX_3\omega^2,$$

that is,

$$\begin{Bmatrix} X_1 \\ X_2 \\ X_3 \end{Bmatrix} = \omega^2 m \begin{bmatrix} 4\alpha_{11} & 2\alpha_{12} & \alpha_{13} \\ 4\alpha_{21} & 2\alpha_{22} & \alpha_{23} \\ 4\alpha_{31} & 2\alpha_{32} & \alpha_{33} \end{bmatrix} \begin{Bmatrix} X_1 \\ X_2 \\ X_3 \end{Bmatrix}.$$

Now,

$$\alpha_{11} = \alpha_{12} = \alpha_{21} = \alpha_{13} = \alpha_{31} = \frac{1}{3k},$$

$$\alpha_{22} = \alpha_{32} = \alpha_{23} = \frac{4}{3k},$$

and

$$\alpha_{33} = \frac{7}{3k}.$$

Hence,

$$\begin{Bmatrix} X_1 \\ X_2 \\ X_3 \end{Bmatrix} = \frac{\omega^2 m}{3k} \begin{bmatrix} 4 & 2 & 1 \\ 4 & 8 & 4 \\ 4 & 8 & 7 \end{bmatrix} \begin{Bmatrix} X_1 \\ X_2 \\ X_3 \end{Bmatrix}.$$

To start the iteration a reasonable estimate for the first mode is

$$\begin{Bmatrix} 1 \\ 2 \\ 4 \end{Bmatrix};$$

this is inversely proportional to the mass ratio of the bodies.
    Eventually iteration for the first mode gives

$$\begin{Bmatrix} 1.0 \\ 3.2 \\ 4.0 \end{Bmatrix} = \frac{14.4\ m\omega^2}{3k} \begin{Bmatrix} 1.0 \\ 3.18 \\ 4.0 \end{Bmatrix}.$$

or $\omega_1 = 0.46\sqrt{(k/m)}$ rad/s.
    To obtain the second principal mode, use the orthogonality relation to remove the first mode from the system matrix:

$$m_1 A_1 A_2 + m_2 B_1 B_2 + m_3 C_1 C_2 = 0.$$

Thus

$$4m(1.0)A_2 + 2m(3.18)B_2 + m(4.0)C_2 = 0,$$

or $A_2 = -1.59B_2 - C_2$, since the first mode is

$$\begin{Bmatrix} 1.0 \\ 3.18 \\ 4.0 \end{Bmatrix}.$$

Hence, rounding 1.59 up to 1.6,

$$\begin{Bmatrix} A_2 \\ B_2 \\ C_2 \end{Bmatrix} = \begin{bmatrix} 0 & -1.6 & -1 \\ 0 & 1 & 0 \\ 0 & 0 & 1 \end{bmatrix} \begin{Bmatrix} A_2 \\ B_2 \\ C_2 \end{Bmatrix}.$$

When this sweeping matrix is combined with the original matrix equation, iteration makes convergence to the second mode take place because the first mode is swept out. Thus,

$$\begin{Bmatrix} X_1 \\ X_2 \\ X_3 \end{Bmatrix} = \frac{\omega^2 m}{3k} \begin{bmatrix} 4 & 2 & 1 \\ 4 & 8 & 4 \\ 4 & 8 & 7 \end{bmatrix} \begin{bmatrix} 0 & -1.6 & -1 \\ 0 & 1 & 0 \\ 0 & 0 & 1 \end{bmatrix} \begin{Bmatrix} X_1 \\ X_2 \\ X_3 \end{Bmatrix}$$

$$= \frac{\omega^2 m}{3k} \begin{bmatrix} 0 & -4.4 & -3 \\ 0 & 1.6 & 0 \\ 0 & 1.6 & 3 \end{bmatrix} \begin{Bmatrix} X_1 \\ X_2 \\ X_3 \end{Bmatrix}.$$

Now estimate the second mode as

$$\begin{Bmatrix} 1 \\ 0 \\ -1 \end{Bmatrix}$$

and iterate:

$$\begin{Bmatrix} 1 \\ 0 \\ -1 \end{Bmatrix} = \frac{\omega^2 m}{3k} \begin{bmatrix} 0 & -4.4 & -3 \\ 0 & 1.6 & 0 \\ 0 & 1.6 & 3 \end{bmatrix} \begin{Bmatrix} 1 \\ 0 \\ -1 \end{Bmatrix} = \frac{m\omega^2}{k} \begin{Bmatrix} 1 \\ 0 \\ -1 \end{Bmatrix}.$$

Hence $\omega_2 = \sqrt{(k/m)}$ rad/s, and the second mode was evidently estimated correctly as $1:0:-1$.

To obtain the third mode, write the orthogonality relation as

$$m_1 A_2 A_3 + m_2 B_2 B_3 + m_3 C_2 C_3 = 0$$

and

$$m_1 A_1 A_3 + m_2 B_1 B_3 + m_3 C_1 C_3 = 0.$$

Substitute

$$A_1 = 1.0, \ B_1 = 3.18, \ C_1 = 4.0$$

and

$$A_2 = 1.0, \ B_2 = 0, \ C_2 = -1.0,$$

as found above. Hence

$$
\begin{Bmatrix} A_3 \\ B_3 \\ C_3 \end{Bmatrix} = \begin{bmatrix} 0 & 0 & 0.25 \\ 0 & 0 & -0.78 \\ 0 & 0 & 1 \end{bmatrix} \begin{Bmatrix} A_3 \\ B_3 \\ C_3 \end{Bmatrix}.
$$

When this sweeping matrix is combined with the equation for the second mode the second mode is removed, so that it yields the third mode on iteration:

$$
\begin{Bmatrix} X_1 \\ X_2 \\ X_3 \end{Bmatrix} = \frac{\omega^2 m}{3k} \begin{bmatrix} 0 & -4.4 & -3 \\ 0 & 1.6 & 0 \\ 0 & 1.6 & 3 \end{bmatrix} \begin{bmatrix} 0 & 0 & 0.25 \\ 0 & 0 & -0.78 \\ 0 & 0 & 1 \end{bmatrix} \begin{Bmatrix} X_1 \\ X_2 \\ X_3 \end{Bmatrix}
$$

$$
= \frac{\omega^2 m}{3k} \begin{bmatrix} 0 & 0 & 0.43 \\ 0 & 0 & -1.25 \\ 0 & 0 & 1.75 \end{bmatrix} \begin{Bmatrix} X_1 \\ X_2 \\ X_3 \end{Bmatrix}.
$$

or

$$
\begin{Bmatrix} X_1 \\ X_2 \\ X_3 \end{Bmatrix} = 1.75 \frac{\omega^2 m}{3k} \begin{bmatrix} 0 & 0 & 0.25 \\ 0 & 0 & -0.72 \\ 0 & 0 & 1 \end{bmatrix} \begin{Bmatrix} X_1 \\ X_2 \\ X_3 \end{Bmatrix}.
$$

An estimate for the third mode shape now has to be made and the iteration procedure carried out once more. In this way the third mode eigenvector is found to be

$$
\begin{Bmatrix} 0.25 \\ -0.72 \\ 1.0 \end{Bmatrix},
$$

and $\omega_3 = 1.32\sqrt{(k/m)}$ rad/s.

The convergence for higher modes becomes more critical if impurities and rounding-off errors are introduced by using the sweeping matrices. One does well to check the highest mode by the inversion of the original matrix equation, which should be equal to the equation formulated in terms of the stiffness influence coefficients.

3.2.1.2 Dunkerley's method

In those cases where it is required to find the lowest, or fundamental natural frequency of free vibration of a multi-degree of freedom system, Dunkerley's method can be used. This is an approximate method which enables a wide range of vibration problems to be solved by using a hand calculator. The method can be understood by considering a two degree of freedom system.

The equations of motion for a two degree of freedom system written in terms of the influence coefficients are

$$y_1 = \alpha_{11}\, m_1\, \omega^2 y_1 + \alpha_{12}\, m_2\, \omega^2\, y_2$$

and

$$y_2 = \alpha_{21}\, m_1\, \omega^2\, y_1 + \alpha_{22}\, m_2\, \omega^2\, y_2$$

so that the frequency equation is given by

$$\begin{vmatrix} \alpha_{11}\, m_1\, \omega^2 - 1 & \alpha_{12}\, m_2\, \omega^2 \\ \alpha_{21}\, m_1\, \omega^2 & \alpha_{22}\, m_2\, \omega^2 - 1 \end{vmatrix} = 0.$$

By expanding this determinant, and solving the resulting quadratic equation, it is found that:

$$\omega_{1,2}^2 = \frac{(\alpha_{11}m_1 + \alpha_{22}m_2) \pm \sqrt{[(\alpha_{11}m_1 + \alpha_{22}m_2)^2 - 4(\alpha_{11}\alpha_{22} - \alpha_{21}\alpha_{12})]}}{2(\alpha_{11}\alpha_{22} - \alpha_{21}\alpha_{12})}$$

Hence,

$$\frac{1}{\omega_1^2} + \frac{1}{\omega_2^2} = 2(\alpha_{11}\alpha_{22} - \alpha_{21}\alpha_{12})$$

$$\times \left[\frac{1}{(\alpha_{11}m_1 + \alpha_{22}m_2) + \sqrt{[(\alpha_{11}m_1 + \alpha_{22}m_2)^2 - 4(\alpha_{11}\alpha_{22} - \alpha_{21}\alpha_{12})]}} \right.$$

$$\left. + \frac{1}{(\alpha_{11}m_1 + \alpha_{22}m_2) - \sqrt{[(\alpha_{11}m_1 + \alpha_{22}m_2)^2 - 4(\alpha_{11}\alpha_{22} - \alpha_{21}\alpha_{12})]}} \right]$$

$$= 2(\alpha_{11}\alpha_{22} - \alpha_{21}\alpha_{12})$$

$$\times \left[\frac{2(\alpha_{11}m_1 + \alpha_{22}m_2)}{(\alpha_{11}m_1 + \alpha_{22}m_2)^2 - (\alpha_{11}m_1 + \alpha_{22}m_2)^2 + 4(\alpha_{11}\alpha_{22} - \alpha_{21}\alpha_{12})} \right],$$

that is,

$$\frac{1}{\omega_1^2} + \frac{1}{\omega_2^2} = \alpha_{11}m_1 + \alpha_{22}m_2.$$

Similarly for an n degree of freedom system,

$$\frac{1}{\omega_1^2} + \frac{1}{\omega_2^2} + \frac{1}{\omega_3^2} + \cdots + \frac{1}{\omega_n^2} = \alpha_{11}m_1 + \alpha_{22}m_2 + \alpha_{33}m_3 + \cdots + \alpha_{nn}m_n.$$

Returning to the analysis for the two degree of freedom system, if P_1 is the natural frequency of body 1 acting alone, then

$$P_1^2 = \frac{k_1}{m_1} = \frac{1}{\alpha_{11}m_1}.$$

Similarly

$$P_2^2 = \frac{1}{\alpha_{22}m_2}.$$

Thus

$$\frac{1}{\omega_1^2} + \frac{1}{\omega_2^2} = \frac{1}{P_1^2} + \frac{1}{P_2^2}.$$

Similarly for an n degree of freedom system,

$$\frac{1}{\omega_1^2} + \frac{1}{\omega_2^2} + \frac{1}{\omega_3^2} + \cdots + \frac{1}{\omega_n^2} = \frac{1}{P_1^2} + \frac{1}{P_2^2} + \frac{1}{P_3^2} + \cdots + \frac{1}{P_n^2}.$$

Usually $\omega_n \gg \cdots \, \omega_3 \gg \omega_2 \gg \omega_1$, so that the LHS of the equation is approximately $\frac{1}{\omega_1^2}$; hence

$$\frac{1}{\omega_1^2} \simeq \frac{1}{P_1^2} + \frac{1}{P_2^2} + \cdots$$

and

$$\frac{1}{\omega_1^2} = \alpha_{11}m_1 + \alpha_{22}m_2 + \cdots.$$

This is referred to as Dunkerley's method for finding the lowest natural frequency of a multi-degree of freedom dynamic system or structure.

Example 23

A three-floor building is modelled by the shear frame shown. Find the frequency of the fundamental mode of free vibration in the plane of the figure if the foundation is capable of translation; m_0 is the effective mass of the foundation, k_0 the transverse stiffness; m_1, m_2 and m_3 are the masses of each floor together with an allowance for the mass of the walls, so that the masses of the walls themselves can be assumed to be zero. The height of each floor is h_1, h_2 and h_3, and the second moment of area of each wall is I_1, I_2 and I_3; the modulus is E.

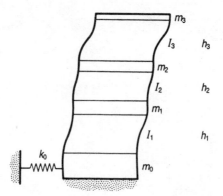

For one wall subjected to a lateral force F at $y = h$ as shown,

$$EI \frac{d^3 x}{dy^3} = -F.$$

Now when $y = 0$ and h, $x = 0$ and $\dfrac{dx}{dy} = 0$, so integrating and applying these boundary conditions gives

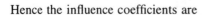

$$x_h = \frac{Fh^3}{12\ EI}.$$

Hence the influence coefficients are

$$\alpha_{00} = \frac{1}{k_0},$$

$$\alpha_{11} = \frac{1}{k_0} + \frac{h_1^3}{24\ EI_1},$$

$$\alpha_{22} = \frac{1}{k_0} + \frac{h_1^3}{24\ EI_1} + \frac{h_2^3}{24\ EI_2},$$

and

$$\alpha_{33} = \frac{1}{k_0} + \frac{h_1^3}{24\ EI_1} + \frac{h_2^3}{24\ EI_2} + \frac{h_3^3}{24\ EI_3}.$$

Applying Dunkerley's method,

$$\frac{1}{\omega_1^2} \simeq \alpha_{00}m_0 + \alpha_{11}m_1 + \alpha_{22}m_2 + \alpha_{33}m_3.$$

If it is assumed that $h_1 = h_2 = h_3 = h$, $I_1 = I_2 = I_3 = I$ and $m_1 = m_2 = m_3 = m$,

$$\frac{1}{\omega_1^2} \simeq \frac{m_0}{k_0} + m\left(\frac{1}{k_0} + \frac{h^3}{24\ EI}\right) + m\left(\frac{1}{k_0} + \frac{2h^3}{24\ EI}\right) + m\left(\frac{1}{k_0} + \frac{3h^3}{24\ EI}\right)$$

$$= \frac{m_0 + 3m}{k_0} + \frac{mh^3}{4EI}.$$

In a particular case, $m_0 = 2 \times 10^6$ kg, $m = 200 \times 10^3$ kg, $h = 4$ m, $E = 200 \times 10^9$ N/m^2, $I = 25 \times 10^{-6}$ m^4 and $k_0 = 10^7$ N/m.
Thus

$$\frac{1}{\omega_1^2} = \frac{(2 \times 10^6) + (3 \times 200 \times 10^3)}{10^7} + \frac{200 \times 10^3 \times 4^3}{4 \times 200 \times 10^9 \times 25 \times 10^{-6}}$$

$$= 0.26 + 0.64,$$

so that $\omega_1 = 1.05$ rad/s, $f_1 = 0.168$ Hz and the period of the oscillation is 5.96 s.

3.2.2 The Lagrange equation

Consideration of the energy in a dynamic system together with the use of the Lagrange equation is a very powerful method of analysis for certain physically complex systems. This is an energy method that allows the equations of motion to be written in terms of any set of *generalized coordinates*. Generalized coordinates are a set of independent parameters that completely specify the system location and that are independent of any constraints. The fundamental form of Lagrange's equation can be written in terms of the generalized coordinates q_i as follows:

$$\frac{d}{dt}\left(\frac{\partial(T)}{\partial \dot{q}_i}\right) - \frac{\partial(T)}{\partial q_i} + \frac{\partial(V)}{\partial q_i} + \frac{\partial(DE)}{\partial \dot{q}_i} = Q_i,$$

where T is the total kinetic energy of the system, V is the total potential energy of the system, DE is the energy dissipation function when the damping is linear (it is half the rate at which energy is dissipated so that for viscous damping $DE = \frac{1}{2}c\dot{x}^2$), Q_i is a generalized external force (or non-linear damping force) acting on the system, and q_i is a generalized coordinate that describes the position of the system.

The subscript i denotes n equations for an n degree of freedom system, so that the Lagrange equation yields as many equations of motion as there are degrees of freedom.

For a free conservative system Q_i and DE are both zero, so that

$$\frac{d}{dt}\left(\frac{\partial(T)}{\partial \dot{q}_i}\right) - \frac{\partial(T)}{\partial q_i} + \frac{\partial(V)}{\partial q_i} = 0.$$

The full derivation of the Lagrange equation can be found in *Vibration Theory and Applications* by W. T. Thomson (Allen & Unwin, 1989).

Example 24

A solid cylinder has a mass M and radius R. Pinned to the axis of the cylinder is an arm of length l which carries a bob of mass m. Obtain the natural frequency of free vibration of the bob. The cylinder is free to roll on the fixed horizontal surface shown.

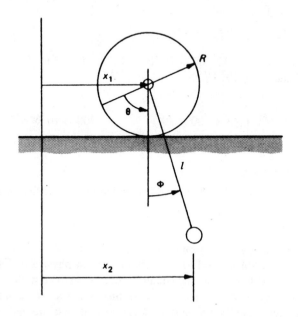

The generalized coordinates are x_1 and x_2. They completely specify the position of the system and are independent of any constraints.

$$T = \tfrac{1}{2}M\dot{x}_1^2 + \tfrac{1}{2}(\tfrac{1}{2}MR^2)\dot{\theta}^2 + \tfrac{1}{2}m\dot{x}_2^2$$

$$= \tfrac{1}{2}M\dot{x}_1^2 + \tfrac{1}{2}(\tfrac{1}{2}M\dot{x}_1^2) + \tfrac{1}{2}m\dot{x}_2^2.$$

$$V = mgl(1 - \cos \phi) = (mgl/2)\phi^2 = (mg/2l)(x_2 - x_1)^2,$$

for small values of ϕ. Apply the Lagrange equation with $q_i = x_1$:

$$(d/dt)(\partial T/\partial \dot{x}_1) = M\ddot{x}_1 + \tfrac{1}{2}M\ddot{x}_1$$

$$\partial V/\partial x_1 = (mg/2l)(-2x_2 + 2x_1).$$

Hence $\tfrac{3}{2}M\ddot{x}_1 + (mg/l)(x_1 - x_2) = 0$ is an equation of motion.

Apply the Lagrange equation with $q_i = x_2$:

$$(d/dt)(\partial T/\partial \dot{x}_2) = m\ddot{x}_2$$

$$\partial V/\partial x_2 = (mg/2l)(2x_2 - 2x_1).$$

Hence $m\ddot{x}_2 + (mg/l)(x_2 - x_1) = 0$ is an equation of motion.

These equations of motion can be solved by assuming that $x_1 = X_1 \sin \omega t$ and $x_2 = X_2 \sin \omega t$. Then

$$X_1((mg/l) - (3M/2)\omega^2) + X_2(-mg/l) = 0$$

and

$$X_1(-mg/l) + X_2((mg/l) - m\omega^2) = 0.$$

The frequency equation is therefore

$$(3M/2)\omega^4 - (g/l)\omega^2(m + (3M/2)) = 0.$$

Thus either $\omega = 0$, or $\omega = \sqrt{((1 + 2m/3M)g/l)}$ rad/s, and $X_1/X_2 = -2m/3M$.

Example 25

To isolate a structure from the vibration generated by a machine, the machine is mounted on a large block. The block is supported on springs as shown. Find the equations that describe the motion of the block in the plane of the figure.

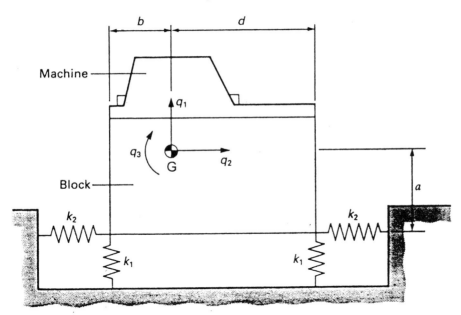

The coordinates used to describe the motion are q_1, q_2 and q_3. These are generalized coordinates because they completely specify the position of the system and are independent of any constraints. If the mass of the block and machine is M, and the total mass moment of inertia about G is I_G, then

$$T = \tfrac{1}{2}M\dot{q}_1^2 + \tfrac{1}{2}M\dot{q}_2^2 + \tfrac{1}{2}I_G\dot{q}_3^2,$$

and

$$V = \text{strain energy stored in the springs}$$

$$= \tfrac{1}{2}k_1(q_1 + bq_3)^2 + \tfrac{1}{2}k_1(q_1 - dq_3)^2 + \tfrac{1}{2}k_2(q_2 + aq_3)^2 + \tfrac{1}{2}k_2(q_2 - aq_3)^2.$$

Now apply the Lagrange equation with $q_i = q_1$.

$$\frac{\partial T}{\partial q_1} = 0.$$

$$\frac{\partial T}{\partial \dot{q}_1} = M\dot{q}_1, \quad \text{so} \quad \frac{d}{dt} \cdot \frac{\partial T}{\dot{q}_1} = M\ddot{q}_1,$$

and

$$\frac{\partial V}{\partial q_1} = k_1(q_1 + bq_3) + k_1(q_1 - dq_3).$$

Thus the first equation of motion is

$$M\ddot{q}_1 + 2k_1q_1 + k_1(b - d)q_3 = 0.$$

Similarly by putting $q_i = q_2$ and $q_i = q_3$, the other equations of motion are obtained as

$$M\ddot{q}_2 + 2k_1q_2 - 2ak_2q_3 = 0$$

and

$$I_G\ddot{q}_3 + k_1(b - d)q_1 - 2ak_2q_2 + (b^2 + d^2)k_1 + 2a^2k_2q_3 = 0.$$

The system therefore has three coordinate-coupled equations of motion. The natural frequencies can be found by substituting $q_i = A_i \sin \omega t$, and solving the resulting frequency equation. It is usually desirable to have all natural frequencies low so that the transmissibility is small throughout the range of frequencies excited.

Example 26

A two-storey building which has its foundation subjected to translation and rotation is modelled by the system shown. Write down expressions for T and V, and indicate how the natural frequencies of free vibration may be found using the Lagrange equation.

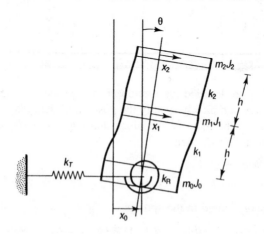

For small-amplitude vibration,

$$T = \tfrac{1}{2}m_0\dot{x}_0^2 + \tfrac{1}{2}J_0\dot{\theta}^2 + \tfrac{1}{2}m_1(\dot{x}_0 + h\dot{\theta} + \dot{x}_1)^2 + \tfrac{1}{2}J_1\dot{\theta}^2$$
$$+ \tfrac{1}{2}m_2(\dot{x}_0 + 2h\dot{\theta} + \dot{x}_2)^2 + \tfrac{1}{2}J_2\dot{\theta}^2$$

and

$$V = \tfrac{1}{2}k_T x_0^2 + \tfrac{1}{2}k_R\theta^2 + 2(\tfrac{1}{2}k_1 x_1^2 + \tfrac{1}{2}k_2(x_2 - x_1)^2)$$

where, x_0, θ, x_1 and x_2 are the generalized coordinates. Substituting in the Lagrange equation with each coordinate gives four equations of motion to be solved for the frequency equation and hence the natural frequencies of free vibration and their associated mode shapes.

3.2.3 Receptance analysis

Some simplification in the analysis of multi-degree of freedom undamped dynamic systems can often be gained by using receptances, particularly if only the natural frequencies are required. If a *harmonic force F* sin vt acts at some point in a system so that the system responds at frequency v, and the point of application of the force has a displacement $x = X$ sin vt, then if the equations of motion are linear, $x = \alpha F$ sin vt, where α, which is a function of the system parameters and v, but not a function of F, is known as the direct receptance at x. If the displacement is determined at some point other than that at which the force is applied, α is known as the transfer or cross receptance.

The analogy with influence coefficients (section 3.2.1) is obvious.

It can be seen that the frequency at which a receptance becomes infinite is a natural frequency of the system. Receptances can be written for rotational and translational coordinates in a system, that is, the slope and deflection at a point.

Thus, if a body of mass m is subjected to a force F sin vt and the response of the body is $x = X$ sin vt,

$$F \sin vt = m\ddot{x} = m(-Xv^2 \sin vt) = -mv^2 x.$$

Thus

$$x = -\frac{1}{mv^2} F \sin vt$$

and

$$\alpha = -\frac{1}{mv^2}.$$

This is the direct receptance of a rigid body.

For a spring, $\alpha = 1/k$. This is the direct receptance of a spring.

In an undamped single degree of freedom model of a system, the equation of motion is

$$m\ddot{x} + kx = F \sin vt.$$

If $x = X$ sin vt, $\alpha = 1/(k - mv^2)$. This is the direct receptance of a single degree of freedom system.

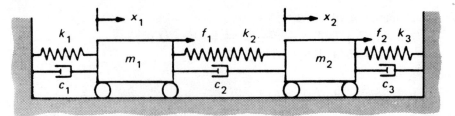

Fig. 3.9.   Two degree of freedom system with forced excitation.

In more complicated systems, it is necessary to be able to distinguish between direct and cross receptances and to specify the points at which the receptances are calculated. This is done by using subscripts. The first subscript indicates the coordinate at which the response is measured, and the second indicates that at which the force is applied. Thus α_{pq}, which is a cross receptance, is the response at p divided by the harmonic force applied at q, and α_{pp} and α_{qq} are direct receptances at p and q respectively.

Consider the two degree of freedom system shown in Fig. 3.9. The equations of motion are

$$m_1\ddot{x}_1 + (k_1 + k_2)x_1 - k_2x_2 = f_1$$

and

$$m_2\ddot{x}_2 + (k_2 + k_3)x_2 - k_2x_1 = 0.$$

Let $f_1 = F_1 \sin vt$, and assume that $x_1 = X_1 \sin vt$ and $x_2 = X_2 \sin vt$. Substituting into the equations of motion gives

$$(k_1 + k_2 - m_1v^2)X_1 + (-k_2)X_2 = F_1$$

and

$$(-k_2)X_1 + (k_2 + k_3 - m_2v^2)X_2 = 0.$$

Thus

$$\alpha_{11} = \frac{X_1}{F_1} = \frac{k_2 + k_3 - m_2v^2}{\Delta},$$

where

$$\Delta = (k_1 + k_2 - m_1v^2)(k_2 + k_3 - m_2v^2) - k_2^2,$$

α_{11} is a direct receptance, and $\Delta = 0$ is the frequency equation.

Also the cross receptance

$$\alpha_{21} = \frac{X_2}{F_1} = \frac{k_2}{\Delta}.$$

This system has two more receptances, the responses due to f_2 applied to the second body. Thus α_{12} and α_{22} may be found. It is a fundamental property that $\alpha_{12} = \alpha_{21}$ (principle of reciprocity) so that symmetrical matrices result.

A general statement of the system response is

$$X_1 = \alpha_{11}F_1 + \alpha_{12}F_2$$

and

$$X_2 = \alpha_{21}F_1 + \alpha_{22}F_2,$$

that is,

$$\left\{ \begin{array}{c} X_1 \\ X_2 \end{array} \right\} = \left[\begin{array}{cc} \alpha_{11} & \alpha_{12} \\ \alpha_{21} & \alpha_{22} \end{array} \right] \left\{ \begin{array}{c} F_1 \\ F_2 \end{array} \right\}.$$

Some simplification in the analysis of complex systems can be achieved by considering the complex system to be a number of simple systems (whose receptances are known) linked together by using conditions of compatibility and equilibrium. The method is to break the complex system down into subsystems and analyse each subsystem separately. Each subsystem receptance is found at the point where it is connected to the adjacent subsystem, and all subsystems are 'joined' together, using the conditions of compatibility and equilibrium.

For example, to find the direct receptance γ_{11} of a dynamic system C at a single coordinate x_1 the system is considered as two subsystems A and B, as shown in Fig. 3.10.

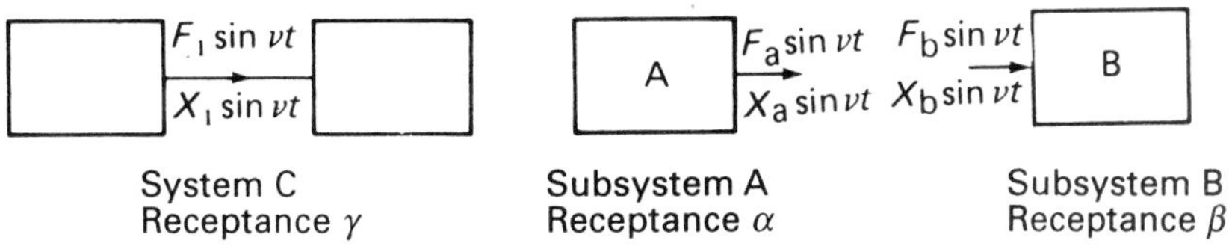

System C
Receptance γ

Subsystem A
Receptance α

Subsystem B
Receptance β

Fig. 3.10.   Dynamic systems.

By definition,

$$\gamma_{11} = \frac{X_1}{F_1}, \quad \alpha_{11} = \frac{X_a}{F_a} \quad \text{and} \quad \beta_{11} = \frac{X_b}{F_b}.$$

Because the systems are connected,

$$X_a = X_b = X_1, \text{ (compatibility)}$$

and

$$F_1 = F_a + F_b, \text{ (equilibrium)}.$$

Hence

$$\frac{1}{\gamma_{11}} = \frac{1}{\alpha_{11}} + \frac{1}{\beta_{11}},$$

that is, the system receptance γ can be found from the receptances of the subsystems.

In a simple spring – body system, subsystems A and B are the spring and body, respectively. Hence $\alpha_{11} = 1/k$, $\beta_{11} = -1/mv^2$ and $1/\gamma_{11} = k - mv^2$, as above.

The frequency equation is $\alpha_{11} + \beta_{11} = 0$, because this condition makes $\gamma_{11} = \infty$.

Example 27

A motor with a flywheel is shown. The motor is to drive a fan of inertia I through a shaft of torsional stiffness k as indicated. The device is required for the ventilation system in a building, and it is necessary to know the fundamental frequency of torsional oscillation so that no unforeseen noise problems will occur. Means for changing this frequency may be sought if it is considered that there may be noise problems. The receptance for torsional oscillation of the motor–flywheel system has been measured at a point P over a limited frequency range which does not include any internal resonances of the system. The figure shows the receptance at P as a function of $(frequency)^2$.

Calculate the lowest non-zero natural frequency for the motor, flywheel and fan system if $I = 0.9$ kg m^2 and $k = 300$ kN m/rad.

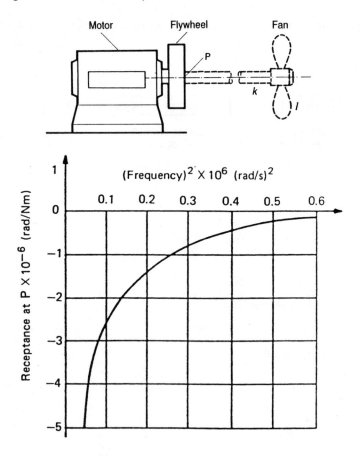

The motor–fan system can be considered to be two subsystems, A and B:

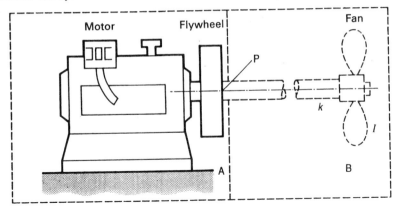

The frequency equation is $\alpha_{11} + \beta_{11} = 0$, where α_{11} is given in graphical form and β_{11} is found as follows. For the fan system,

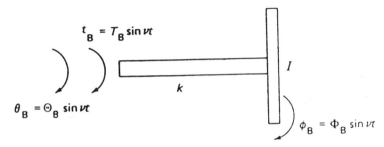

Now

$$t_B = k(\theta_B - \phi_B) = I\ddot{\phi}_B = -I\Phi_B v^2 \sin vt,$$

so

$$k(\Theta_B - \Phi_B) = -I\Phi_B v^2$$

and

$$\Theta_B = \Phi_B - \frac{I}{k}\Phi_B v^2 = \Phi_B\left(\frac{k - Iv^2}{k}\right).$$

Thus

$$\beta_{11} = \frac{\Theta_B}{T_B} = -\Phi_B\left(\frac{k - Iv^2}{k}\right)\frac{1}{\Phi_B I v^2}$$

$$= -\left(\frac{k - Iv^2}{kIv^2}\right).$$

If α_{11} and $-\beta_{11}$ are plotted as functions of (frequency)^2, the intersection gives the value of the frequency that is a solution of $\alpha_{11} + \beta_{11} = 0$, that is, the natural frequency of free vibration is found. The table below can be calculated for $-\beta_{11}$ because $k = 300 \times 10^3$ N m/rad, and $I = 0.9$ kg m^2.

v^2	Iv^2	$k - Iv^2$	kIv^2	$-\beta_{11}$
0.3×10^6	0.27×10^6	0.03×10^6	0.081×10^{12}	-0.37×10^{-6}
0.4×10^6	0.36×10^6	0.06×10^6	0.108×10^{12}	-0.55×10^{-6}
0.5×10^6	0.45×10^6	0.15×10^6	0.135×10^{12}	-1.11×10^{-6}

The receptance β_{11} can now be plotted against (frequency)^2 as below:

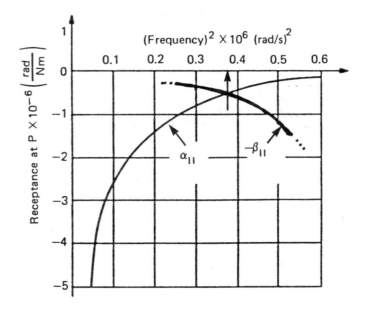

The intersection occurs at (frequency)$^2 = 0.377 \times 10^6$ (rad/s)^2, that is, a frequency of 614 rad/s or 97.7 Hz. This is the natural frequency of the combined motor–fan system. It can be seen that the effect of using different fans with different k and I values is easily found, without having to re-analyse the whole motor–fan system.

Some subsystems, such as those shown in Fig. 3.11, are linked by two coordinates, for example deflection and slope at the common point.

Now in this case,

$$X_{a1} = \alpha_{11}F_{a1} + \alpha_{12}F_{a2},$$

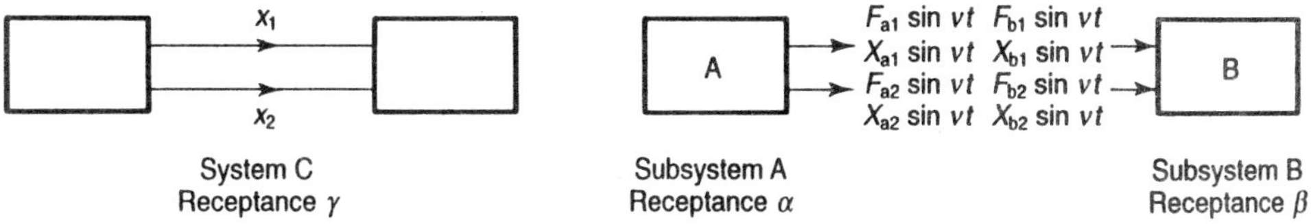

Fig. 3.11.   Applied forces and system responses.

$$X_{a2} = \alpha_{21}F_{a1} + \alpha_{22}F_{a2},$$

$$X_{b1} = \beta_{11}F_{b1} + \beta_{12}F_{b2}$$

and

$$X_{b2} = \beta_{21}F_{b1} + \beta_{22}F_{b2}.$$

The applied forces or moments are $F_1 \sin vt$ and $F_2 \sin vt$ where

$$F_1 = F_{a1} + F_{b1}$$

and

$$F_2 = F_{a2} + F_{b2}.$$

Since the subsystems are linked,

$$X_1 = X_{a1} = X_{b1}$$

and

$$X_2 = X_{a2} = X_{b2}.$$

Hence if excitation is applied at x_1 only, $F_2 = 0$ and

$$\gamma_{11} = \frac{X_1}{F_1} = \frac{\alpha_{11}(\beta_{11}\beta_{22} - \beta_{12})^2 + \beta_{11}(\alpha_{11}\alpha_{22} - \alpha_{12})^2}{\Delta},$$

where

$$\Delta = (\alpha_{11} + \beta_{11})(\alpha_{22} + \beta_{22}) - (\alpha_{12} + \beta_{12})^2$$

and

$$\gamma_{21} = \frac{X_2}{F_1} = \frac{\alpha_{12}(\beta_{11}\beta_{22} - \alpha_{12}\beta_{12}) - \beta_{12}(\alpha_{11}\alpha_{22} - \alpha_{12}\beta_{12})}{\Delta}.$$

If

$$F_1 = 0,$$

$$\gamma_{22} = \frac{X_2}{F_2} = \frac{\alpha_{22}(\beta_{11}\beta_{22} - \beta_{12})^2 - \beta_{22}(\alpha_{11}\alpha_{22} - \alpha_{12})^2}{\Delta}.$$

Since $\Delta = 0$ is the frequency equation, the natural frequencies of the system C are given by

$$\begin{vmatrix} \alpha_{11} + \beta_{11} & \alpha_{12} + \beta_{12} \\ \alpha_{21} + \beta_{21} & \alpha_{22} + \beta_{22} \end{vmatrix} = 0.$$

This is an extremely useful method for finding the frequency equation of a system because only the receptances of the subsystems are required. The receptances of many dynamic systems have been published in *The Mechanics of Vibration* by R. E. D. Bishop and D. C. Johnson (CUP, 1960/1979). By repeated application of this method, a system can be considered to consist of any number of subsystems. This technique is, therefore, ideally suited to a computer solution.

It should be appreciated that although the receptance technique is useful for writing the frequency equation, it does not simplify the solution of this equation.

3.2.4 Impedance and mobility analysis

Impedance and mobility analysis techniques are frequently applied to systems and structures with many degrees of freedom. However, the method is best introduced by considering simple systems initially.

The impedance of a body is the ratio of the amplitude of the *harmonic* exciting force applied, to the amplitude of the resulting velocity. The mobility is the reciprocal of the impedance. It will be appreciated, therefore, that impedance and mobility analysis techniques are similar to those used in the receptance analysis of dynamic systems.

For a body of mass m subjected to a harmonic exciting force represented by $Fe^{j\nu t}$ the resulting motion is $x = Xe^{j\nu t}$. Thus

$$Fe^{j\nu t} = m\ddot{x} = -m\nu^2 Xe^{j\nu t},$$

and the receptance of the body,

$$\frac{X}{F} = -\frac{1}{m\nu^2}.$$

Now

$$Fe^{j\nu t} = -m\nu^2 Xe^{j\nu t}$$
$$= mj\nu(j\nu Xe^{j\nu t}) = mj\nu v,$$

where v is the velocity of the body, and $v = Ve^{j\nu t}$.

Thus the impedance of a body of mass m is Z_m, where

$$Z_m = \frac{F}{V} = jm\nu,$$

and the mobility of a body of mass m is M_m, where

$$M_m = \frac{V}{F} = \frac{1}{jmv}.$$

Putting $s = jv$ so that $x = Xe^{st}$ gives

$$Z_m = ms,$$

$$M_m = \frac{1}{ms}$$

and

$$V = sX.$$

For a spring of stiffness k, $Fe^{jvt} = kXe^{jvt}$ and thus $Z_k = F/V = k/s$ and $M_k = s/k$, whereas for a viscous damper of coefficient c, $Z_c = c$ and $M_c = 1/c$.

If these elements of a dynamic system are combined so that the velocity is common to all elements, then the impedances may be added to give the system impedance, whereas if the force is common to all elements the mobilities may be added. This is demonstrated below by considering a spring–mass single degree of freedom system with viscous damping, as shown in Fig. 3.12.

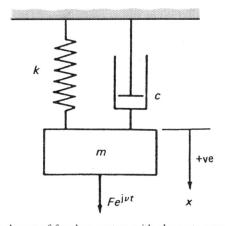

Fig. 3.12.   Single degree of freedom system with elements connected in parallel.

The velocity of the body is common to all elements, so that the force applied is the sum of the forces required for each element. The system impedance,

$$Z = \frac{F}{V} = \frac{F_m + F_k + F_c}{V}$$

$$= Z_m + Z_k + Z_c.$$

Hence

$$Z = ms + \frac{k}{s} + c,$$

that is,

$$F = (ms^2 + cs + k)X$$

or

$$F = (k - mv^2 + jcv)X.$$

Hence

$$X = \frac{F}{\sqrt{[(k - mv^2)^2 + (cv)^2]}}.$$

Thus when system elements are connected in parallel their impedances are added to give the system impedance.

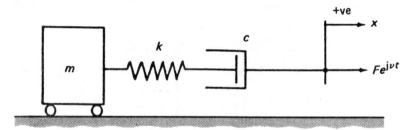

Fig. 3.13.   Single degree of freedom system with elements connected in series.

In the system shown in Fig. 3.13, however, the force is common to all elements. In this case the force on the body is common to all elements so that the velocity at the driving point is the sum of the individual velocities. The system mobility,

$$M = \frac{V}{F} = \frac{V_m + V_k + V_c}{F}$$

$$= M_m + M_k + M_c$$

$$= \frac{1}{ms} + \frac{s}{k} + \frac{1}{c}.$$

Thus when system elements are connected in series their mobilities are added to give the system mobility.

In the system shown in Fig. 3.14, the system comprises a spring and damper connected in series with a body connected in parallel.

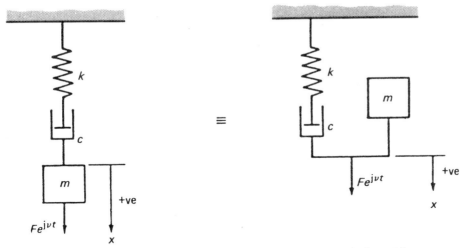

Fig. 3.14.   Single degree of freedom system and impedance analysis model.

Thus the spring and damper mobilities can be added or the reciprocal of their impedances can be added. Hence the system driving point impedance Z is given by

$$Z = Z_m + \left[\frac{1}{Z_k} + \frac{1}{Z_c} \right]^{-1}$$

$$= ms + \left[\frac{1}{k/s} + \frac{1}{c} \right]^{-1}$$

$$= \frac{mcs^2 + mks + kc}{cs + k}.$$

Consider the system shown in Fig. 3.15. The spring k_1 and the body m_1 are connected in parallel with each other and are connected in series with the damper c_1.

Thus the driving point impedance Z is

$$Z = Z_{m_1} + Z_{k_2} + Z_{c_2} + Z_1$$

where

$$Z_1 = \frac{1}{M_1},$$

$$M_1 = M_{c_1} + M_2,$$

$$M_2 = \frac{1}{Z_2},$$

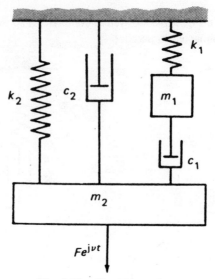

Fig. 3.15.   Dynamic system.

and

$$Z_2 = Z_{k_1} + Z_{m_1}.$$

Thus

$$Z = Z_{m_2} + Z_{k_2} + Z_{c_2} + \cfrac{1}{\cfrac{1}{Z_{c_1}} + \cfrac{1}{Z_{k_1} + Z_{m_1}}}.$$

Hence

$$Z = \frac{\begin{aligned}m_1 m_2 s^4 &+ (m_1 c_2 + m_2 c_1 + m_1 c_1)s^3 \\ &+ (m_1 k_1 + m_2 k_2 + c_1 c_2)s^2 + (c_1 k_2 + c_2 k_1 + c_1 k_1)s + k_1 k_2\end{aligned}}{s(m_1 s^2 + c_1 s + k_1)}.$$

The frequency equation is given when the impedance is made equal to zero or when the mobility is infinite. Thus the natural frequencies of the system can be found by putting $s = j\omega$ in the numerator above and setting it equal to zero.

To summarize, the mobility and impedance of individual elements in a dynamic system are calculated on the basis that the velocity is the relative velocity of the two ends of a spring or a damper, rather than the absolute velocity of the body. Individual impedances are added for elements or subsystems connected in parallel, and individual mobilities are added for elements or subsystems connected in series.

Example 28

Find the driving point impedance of the system shown in Fig. 3.6, and hence obtain the frequency equation.

The system of Fig. 3.6 can be redrawn as shown.

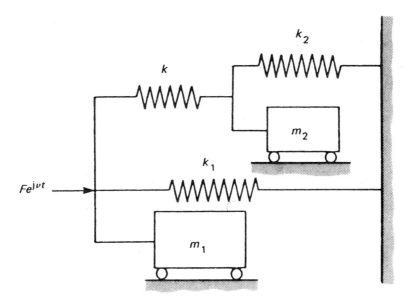

The driving point impedance is therefore

$$Z = Z_{m_1} + Z_{k_1} + \cfrac{1}{\cfrac{1}{Z_k} + \cfrac{1}{Z_{m_2} + Z_{k_2}}}$$

$$= m_1 s + \cfrac{k_1}{s} + \cfrac{1}{\cfrac{1}{k/s} + \cfrac{1}{m_2 s + k_2/s}}$$

$$= \frac{(m_1 s^2 + k_1)(m_2 s^2 + k + k_2) + (m_2 s^2 + k_2)k}{s(m_2 s^2 + k + k_2)}.$$

The frequency equation is obtained by putting $Z = 0$ and $s = j\omega$, thus:

$$(k_1 - m_1 \omega^2)(k + k_2 - m_2 \omega^2) + k(k_2 - m_2 \omega^2) = 0.$$

3.3 MODAL ANALYSIS TECHNIQUES

It is shown in section 3.1.2 that in a dynamic system with coupled coordinates of motion it is possible, under certain conditions, to uncouple the modes of vibration. If this is arranged, the motions expressed by each coordinate can take place independently. These coordinates are then referred to as principal coordinates. This is the basis of the modal analysis technique; that is, independent equations of motion are obtained for each mode of the dynamic response of a multi-degree of freedom system, by uncoupling the differential equations of motion. For each mode of vibration, therefore, there is one independent equation of motion which can be solved as if it were the equation of motion of a single degree of freedom system. The dynamic response of the whole structure or system is then obtained by superposition of the responses of the individual modes. This is usually simpler than simultaneously solving coupled differential equations of motion.

It is often difficult to determine the damping coefficients in the coupled equations of motion analysis, but with the modal analysis method the effect of damping on the system response can be determined by using typical modal damping factors obtained from experiment or previous work. Analytical techniques for predicting structural vibration have become increasingly sophisticated, but the prediction of damping remains difficult. Accordingly very many practical problems related to the vibration of real structures are solved using experimentally based analytical methods.

The availability of powerful mini-computers and Fast Fourier Transform (FFT) analysers have made the acquisition and analysis of experimental data fast, economic and reliable, so that measured data for mass, stiffness and damping properties of each mode of vibration of a structure are readily obtained. To do this, the frequency response function of a structure is usually obtained experimentally by exciting the structure at some point with a measured input force, and measuring the response of the structure at another point. The response is usually measured by an accelerometer (see section 2.3.10). The excitation is often provided by an electromagnetic shaker or by impact, and measured directly with a piezoelectric force gauge. The damping associated with each mode can be found from the FFT analyser data by frequency bandwidth measurement, Nyquist diagrams or curve fit algorithms which estimate modal mass, stiffness and damping from the response curves.

The technique for obtaining modal data from experiment is called modal testing. Some measurement errors can be eliminated to make the data consistent, but a disadvantage of the technique is that modal data is often incomplete and may not be able to represent actual damping accurately. Figure 3.16 shows typical response plots for a structure. Three distinct resonances and modes are evident. The data from these three modes is shown in Fig. 3.17 in terms of modulus and phase of the response over a relevant frequency range.

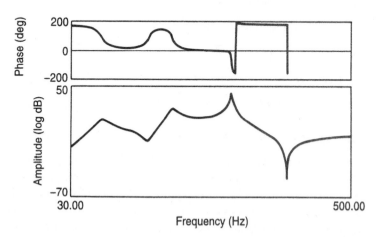

Fig. 3.16.   Typical phase and amplitude versus frequency plots for a structure.

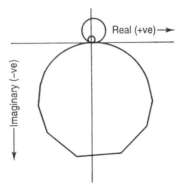

Fig. 3.17.   Modulus and phase plots from data of Fig. 3.16 for three modes of vibration. Frequency range 60–320 Hz.

Modes 1, 2 and 3 are considered in greater detail in Figs 3.18, 3.19 and 3.20, respectively. Analysing the data from mode 1 using a circle fit gives a resonance frequency of 80.1 Hz and a damping loss factor of 0.2. For mode 2, the figures are 200 Hz and 0.05, and for mode 3, 300 Hz and 0.01. These results should be compared with the data shown in Fig. 3.16.

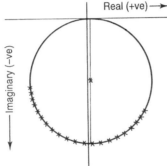

Fig. 3.18.   First mode analysis; $f_1 = 80.1$ Hz, $\eta_1 = 0.20$.

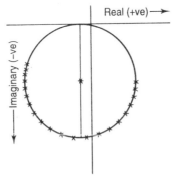

Fig. 3.19.   Second mode analysis; $f_2 = 200$ Hz, $\eta_2 = 0.05$.

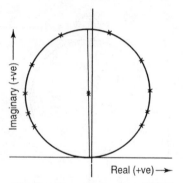

Fig. 3.20.   Third mode analysis; $f_3 = 300$ Hz, $\eta_3 = 0.01$.

Modal analysis is discussed in some detail in *Modal Analysis: Theory and Practice* by D. J. Ewins (Research Studies Press, 1985), and in the Proceedings of the International Modal Analysis Conferences (IMAC) held annually in the USA.

4

The vibration of continuous structures

Continuous structures such as beams, rods, cables and plates can be modelled by discrete mass and stiffness parameters and analysed as multi-degree of freedom systems, but such a model is not sufficiently accurate for most purposes. Furthermore, mass and elasticity cannot always be separated in models of real systems. Thus mass and elasticity have to be considered as distributed or continuous parameters.

For the analysis of structures with distributed mass and elasticity it is necessary to assume a homogeneous, isotropic material that follows Hooke's law.

Generally, free vibration is the sum of the *principal modes*. However, in the unlikely event of the elastic curve of the body in which motion is excited coinciding exactly with one of the principal modes, only that mode will be excited. In most continuous structures the rapid damping out of high-frequency modes often leads to the fundamental mode predominating.

4.1 LONGITUDINAL VIBRATION OF A THIN UNIFORM BEAM

Consider the longitudinal vibration of a thin uniform beam of cross-sectional area S, material density ρ, and modulus E under an axial force P, as shown in Fig. 4.1.

The net force acting on the element is $P + \partial P/\partial x \,.\, \mathrm{d}x - P$, and this is equal to the product of the mass of the element and its acceleration.

From Fig. 4.1,

$$\frac{\partial P}{\partial x}\,\mathrm{d}x = \rho\, S\, \mathrm{d}x\,\frac{\partial^2 u}{\partial t^2}.$$

Now strain $\partial u/\partial x = P/SE$, so

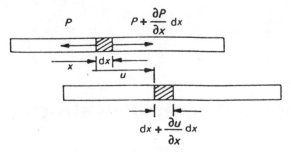

Fig. 4.1. Longitudinal beam vibration.

$$\partial P/\partial x = SE(\partial^2 u/\partial x^2).$$

Thus

$$\partial^2 u/\partial t^2 = (E/\rho)(\partial^2 u/\partial x^2),$$

or

$$\partial^2 u/\partial x^2 = (1/c^2)(\partial^2 u/\partial t^2), \quad \text{where } c = \sqrt{(E/\rho)}.$$

This is the wave equation. The velocity of propagation of the displacement or stress wave in the bar is c.

The wave equation

$$\frac{\partial^2 u}{\partial x^2} = \left(\frac{1}{c^2}\right)\left(\frac{\partial^2 u}{\partial t^2}\right)$$

can be solved by the method of separation of variables and assuming a solution of the form

$$u(x, t) = F(x)G(t).$$

Substituting this solution into the wave equation gives

$$\frac{\partial^2 F(x)}{\partial x^2} G(t) = \frac{1}{c^2} \frac{\partial^2 G(t)}{\partial t^2} F(x),$$

that is

$$\frac{1}{F(x)} \frac{\partial^2 F(x)}{\partial x^2} = \frac{1}{c^2} \frac{1}{G(t)} \frac{\partial^2 G(t)}{\partial t^2}.$$

The LHS is a function of x only, and the RHS is a function of t only, so partial derivatives are no longer required. Each side must be a constant, $-(\omega/c)^2$ say. (This quantity is chosen for convenience of solution.) Then

$$\frac{d^2 F(x)}{dx^2} + \left(\frac{\omega}{c}\right)^2 F(x) = 0$$

and

$$\frac{d^2 G(t)}{dt^2} + \omega^2 G(t) = 0.$$

Hence

$$F(x) = A \sin\left(\frac{\omega}{c}\right) x + B \cos\left(\frac{\omega}{c}\right) x$$

and

$$G(t) = C \sin \omega t + D \cos \omega t.$$

The constants A and B depend upon the boundary conditions, and C and D upon the initial conditions. The complete solution to the wave equation is therefore

$$u = \left(A \sin\left(\frac{\omega}{c}\right) x + B \cos\left(\frac{\omega}{c}\right) x\right)\left(C \sin \omega t + D \cos \omega t\right).$$

Example 29

Find the natural frequencies and mode shapes of longitudinal vibrations for a free–free beam with initial displacement zero.

Since the beam has free ends, $\partial u / \partial x = 0$ at $x = 0$ and $x = l$. Now

$$\frac{\partial u}{\partial x} = \left(A\left(\frac{\omega}{c}\right)\cos\left(\frac{\omega}{c}\right) x - B\left(\frac{\omega}{c}\right)\sin\left(\frac{\omega}{c}\right) x\right)\left(C \sin \omega t + D \cos \omega t\right).$$

Hence

$$\left(\frac{\partial u}{\partial x}\right)_{x=0} = A\left(\frac{\omega}{c}\right)(C \sin \omega t + D \cos \omega t) = 0, \quad \text{so that } A = 0$$

and

$$\left(\frac{\partial u}{\partial x}\right)_{x=l} = \left(\frac{\omega}{c}\right)\left(-B \sin\left(\frac{\omega l}{c}\right)\right)\left(C \sin \omega t + D \cos \omega t\right) = 0.$$

Thus $\sin(\omega l/c) = 0$, since $B \neq 0$, and therefore

$$\frac{\omega l}{c} = \frac{\omega l}{\sqrt{(E/\rho)}} = \pi, 2\pi, ..., n\pi, ...,$$

that is,

$$\omega_n = \frac{n\pi}{l}\sqrt{\left(\frac{E}{\rho}\right)} \text{ rad/s},$$

where $\omega = c/$wavelength. These are the natural frequencies.
  If the initial displacement is zero, $D = 0$ and

$$u_n = B' \cos\left(\frac{n\pi}{l}\right)x \cdot \sin\left(\frac{n\pi}{l}\right)\sqrt{\left(\frac{E}{\rho}\right)} t.$$

where $B' = B \times C$. Hence the mode shape is determined.

Example 30

A uniform vertical rod of length l and cross-section S is fixed at the upper end and is loaded with a body of mass M on the other. Show that the natural frequencies of longitudinal vibration are determined by

$$\omega l \sqrt{(\rho/E)} \tan \omega l \sqrt{(\rho/E)} = S\rho l/M.$$

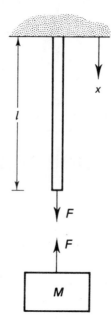

At $x = 0$, $u = 0$, and at $x = l$, $F = SE (\partial u/\partial x)$.

Also

$$F = SE (\partial u/\partial x) = -M(\partial^2 u/\partial t^2).$$

The general solution is

$$u = (A \sin(\omega/c)x + B \cos(\omega/c)x)(C \sin \omega t + D \cos \omega t).$$

Now, $u_{x=0} = 0$, so $B = 0$,

thus

$$u = (A \sin(\omega/c)x)(C \sin \omega t + D \cos \omega t),$$

$$(\partial u/\partial x)_{x=l} = (A(\omega/c) \cos(\omega l/c))(C \sin \omega t + D \cos \omega t)$$

and

$$(\partial^2 u/\partial t^2)_{x=l} = (-A\omega^2 \sin(\omega l/c))(C \sin \omega t + D \cos \omega t),$$

so

$$F = SEA (\omega/c) \cos(\omega l/c)(C \sin \omega t + D \cos \omega t)$$

$$= MA\omega^2 \sin(\omega l/c)(C \sin \omega t + D \cos \omega t).$$

Hence $(\omega l/c) \tan(\omega l/c) = SlE/Mc^2$, and

$$\omega l\sqrt{(\rho/E)} \tan \omega l\sqrt{(\rho/E)} = S\rho l/M, \quad \text{since } c^2 = E/\rho.$$

4.2 TRANSVERSE VIBRATION OF A THIN UNIFORM BEAM

The transverse or lateral vibration of a thin uniform beam is another vibration problem in which both elasticity and mass are distributed. Consider the moments and forces acting on the element of the beam shown in Fig. 4.2. The beam has a cross-sectional area A, flexural rigidity EI, material of density ρ and Q is the shear force.

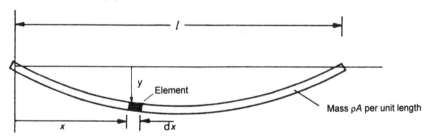

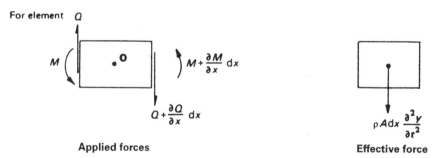

Fig. 4.2.   Transverse beam vibration.

Then for the element, neglecting rotary inertia and shear of the element, taking moments about O gives

$$M + Q\frac{dx}{2} + Q\frac{dx}{2} + \frac{\partial Q}{\partial x} dx \frac{dx}{2} = M + \frac{\partial M}{\partial x} dx,$$

that is,

$$Q = \partial M/\partial x.$$

Summing forces in the y direction gives

$$\frac{\partial Q}{\partial x}\, dx = \rho A\, dx\, \frac{\partial^2 y}{\partial t^2}.$$

Hence

$$\frac{\partial^2 M}{\partial x^2} = \rho A\, \frac{\partial^2 y}{\partial t^2}.$$

Now EI is a constant for a prismatical beam, so

$$M = -EI\, \frac{\partial^2 y}{\partial x^2} \quad \text{and} \quad \frac{\partial^2 M}{\partial x^2} = -EI\, \frac{\partial^4 y}{\partial x^4}.$$

Thus

$$\frac{\partial^4 y}{\partial x^4} + \left(\frac{\rho A}{EI}\right) \frac{\partial^2 y}{\partial t^2} = 0.$$

This is the general equation for the transverse vibration of a uniform beam.

When a beam performs a normal mode of vibration the deflection at any point of the beam varies harmonically with time, and can be written

$$y = X\, (B_1 \sin \omega t + B_2 \cos \omega t),$$

where X is a function of x which defines the beam shape of the normal mode of vibration. Hence

$$\frac{d^4 X}{dx^4} = \left(\frac{\rho A}{EI}\right) \omega^2 X = \lambda^4 X,$$

where

$\lambda^4 = \rho A \omega^2/EI$.   This is the *beam equation.*

The general solution to the beam equation is

$$X = C_1 \cos \lambda x + C_2 \sin \lambda x + C_3 \cosh \lambda x + C_4 \sinh \lambda x,$$

where the constants $C_{1,2,3,4}$ are determined from the boundary conditions.

For example, consider the transverse vibration of a thin prismatical beam of length l, simply supported at each end. The deflection and bending moment are therefore zero at each end, so that the boundary conditions are $X = 0$ and $d^2X/dx^2 = 0$ at $x = 0$ and $x = l$.

Substituting these boundary conditions into the general solution above gives

at $x = 0$, $X = 0$;   thus $0 = C_1 + C_3$,

and

$$\text{at } x = 0, \quad \frac{d^2X}{dx^2} = 0; \quad \text{thus } 0 = C_1 - C_3;$$

that is,

$$C_1 = C_3 = 0 \quad \text{and} \quad X = C_2 \sin \lambda x + C_4 \sinh \lambda x.$$

Now

$$\text{at } x = l, X = 0 \quad \text{so that} \quad 0 = C_2 \sin \lambda l + C_4 \sinh \lambda l,$$

and

$$\text{at } x = l, \quad \frac{d^2X}{dx^2} = 0, \quad \text{so that} \quad 0 = C_2 \sin \lambda l - C_4 \sinh \lambda l;$$

that is,

$$C_2 \sin \lambda l = C_4 \sinh \lambda l = 0.$$

Since $\lambda l \neq 0$, $\sinh \lambda l \neq 0$ and therefore $C_4 = 0$.

Also $C_2 \sin \lambda l = 0$. Since $C_2 \neq 0$ otherwise $X = 0$ for all x, then $\sin \lambda l = 0$. Hence $X = C_2 \sin \lambda x$ and the solutions to $\sin \lambda l = 0$ give the natural frequencies. These are

$$\lambda = 0, \frac{\pi}{l}, \frac{2\pi}{l}, \frac{3\pi}{l}, \dots$$

so that

$$\omega = 0, \left(\frac{\pi}{l}\right)^2 \sqrt{\left(\frac{EI}{A\rho}\right)}, \left(\frac{2\pi}{l}\right)^2 \sqrt{\left(\frac{EI}{A\rho}\right)}, \left(\frac{3\pi}{l}\right)^2 \sqrt{\left(\frac{EI}{A\rho}\right)}, \dots \text{rad/s};$$

$\lambda = 0$, $\omega = 0$ is a trivial solution because the beam is at rest, so the lowest or first natural frequency is $\omega_1 = (\pi/l)^2\sqrt{(EI/A\rho)}$ rad/s, and the corresponding mode shape is $X = C_2 \sin \pi x/l$; this is the first mode; $\omega_2 = (2\pi/l)^2\sqrt{(EI/A\rho)}$ rad/s is the second natural frequency, and the second mode is $X = C_2 \sin 2\pi x/l$, and so on. The mode shapes are drawn in Fig. 4.3.

These sinusoidal vibrations can be superimposed so that any initial conditions can be represented. Other end conditions give frequency equations with the solution where the values of α are given in Table 4.1.

$$\omega = \frac{\alpha}{l^2} \sqrt{\left(\frac{EI}{A\rho}\right)} \text{ rad/s},$$

1st mode shape, one half-wave:

$$y = C_2 \sin \pi \left(\frac{x}{l}\right)(B_1 \sin \omega_1 t + B_2 \cos \omega_1 t); \quad \omega_1 = \left(\frac{\pi}{l}\right)^2 \sqrt{\left(\frac{EI}{A\rho}\right)} \text{ rad/s.}$$

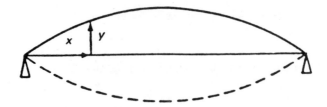

2nd mode shape, two half-waves:

$$y = C_2 \sin 2\pi \left(\frac{x}{l}\right)(B_1 \sin \omega_2 t + B_2 \cos \omega_2 t); \quad \omega_2 = \left(\frac{2\pi}{l}\right)^2 \sqrt{\left(\frac{EI}{A\rho}\right)} \text{ rad/s.}$$

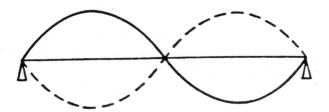

3rd mode shape, three half-waves:

$$y = C_2 \sin 3\pi \left(\frac{x}{l}\right)(B_1 \sin \omega_3 t + B_2 \cos \omega_3 t); \quad \omega_3 = \left(\frac{3\pi}{l}\right)^2 \sqrt{\left(\frac{EI}{A\rho}\right)} \text{ rad/s.}$$

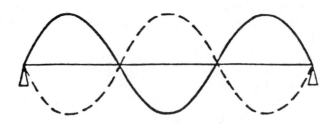

Fig. 4.3. Transverse beam vibration mode shapes and frequencies.

Table 4.1

End conditions	Frequency equation	1st mode	2nd mode	3rd mode	4th mode	5th mode
Clamped–free	$\cos \lambda l \cosh \lambda l = -1$	3.52	22.4	61.7	21.0	199.9
Pinned–pinned	$\sin \lambda l = 0$	9.87	39.5	88.9	157.9	246.8
Clamped–pinned	$\tan \lambda l = \tanh \lambda l$	15.4	50.0	104.0	178.3	272.0
Clamped–clamped or Free-free	$\cos \lambda l \cosh \lambda l = 1$	22.4	61.7	121.0	199.9	298.6

The natural frequencies and mode shapes of a wide range of beams and structures are given in *Formulas for Natural Frequency and Mode Shape* by R. D. Blevins (Van Nostrand, 1979).

4.2.1 The whirling of shafts

An important application of the theory for transverse beam vibration is to the whirling of shafts. If the speed of rotation of a shaft is increased, certain speeds will be reached at which violent instability occurs. These are the critical speeds of whirling. Since the loading on the shaft is due to centrifugal effects the equation of motion is exactly the same as for transverse beam vibration. The centrifugal effects occur because it is impossible to make the centre of mass of any section coincide exactly with the axis of rotation, because of a lack of homogeneity in the material and other practical difficulties.

Example 31

A uniform steel shaft which is carried in long bearings at each end has an effective unsupported length of 3 m. Calculate the first two whirling speeds.

Take $I/A = 0.1 \times 10^{-3}$ m^2, $E = 200$ GN/m^2, and $\rho = 8000$ kg/m^3.

Since the shaft is supported in long bearings, it can be considered to be 'built in' at each end so that, from Table 4.1,

$$\omega = \frac{\alpha_1}{l^2}\sqrt{\left|\left(\frac{EI}{A\rho}\right)\right|} \text{ rad/s},$$

where $\alpha_1 = 22.4$ and $\alpha_2 = 61.7$. For the shaft,

$$\sqrt{\left|\left(\frac{EI}{A\rho}\right)\right|} = \sqrt{\left|\left(\frac{200 \times 10^9 \times 0.1 \times 10^{-3}}{8000}\right)\right|} = 50 \text{ m}^2/\text{s},$$

so that the first two whirling speeds are:

$$\omega_1 = \frac{22.4}{9} \ 50 = 124.4 \text{ rad/s},$$

so

$$f_1 = \frac{\omega_1}{2\pi} = \frac{124.4}{2\pi} = 19.8 \text{ cycle/s and } N_1 = 1188 \text{ rev/min}$$

and

$$N_2 = \frac{61.7}{22.4} \, 1188 = 3272 \text{ rev/min}.$$

Rotating this shaft at speeds at or near to the above will excite severe resonance vibration.

4.2.2 Rotary inertia and shear effects

When a beam is subjected to lateral vibration so that the depth of the beam is a significant proportion of the distance between two adjacent nodes, rotary inertia of beam elements and transverse shear deformation arising from the severe contortions of the beam during vibration make significant contributions to the lateral deflection. Therefore rotary inertia and shear effects must be taken into account in the analysis of high-frequency vibration of all beams, and in all analyses of deep beams.

The moment equation can be modified to take into account rotary inertia by a term $\rho I \, \partial^3 y/(\partial x \, \partial t^2)$, so that the beam equation becomes

$$EI \frac{\partial^4 y}{\partial x^4} - \rho I \frac{\partial^3 y}{\partial x \partial t^2} + \rho A \frac{\partial^2 y}{\partial t^2} = 0.$$

Shear deformation effects can be included by adding a term

$$\frac{EI\rho}{kg} \frac{\partial^4 y}{\partial x^2 \partial t^2},$$

where k is a constant whose value depends upon the cross section of the beam. Generally, k is about 0.85. The beam equation then becomes

$$EI \frac{\partial^4 y}{\partial x^4} - \frac{EI\rho}{kg} \frac{\partial^4 y}{\partial x^2 \partial t^2} + \rho A \frac{\partial^2 y}{\partial t^2} = 0.$$

Solutions to these equations are available, which generally lead to a frequency a few percent more accurate than the solution to the simple beam equation. However, in most cases the modelling errors exceed this. In general, the correction due to shear is larger than the correction due to rotary inertia.

4.2.3 The effect of axial loading

Beams are often subjected to an axial load, and this can have a significant effect on the lateral vibration of the beam. If an axial tension T exists, which is assumed to be constant for small-amplitude beam vibrations, the moment equation can be modified by including a term $T\partial^2 y/\partial x^2$, so that the beam equation becomes

$$EI\frac{\partial^4 y}{\partial x^4} - T\frac{\partial^2 y}{\partial x^2} + \rho A\frac{\partial^2 y}{\partial t^2} = 0.$$

Tension in a beam will increase its stiffness and therefore increase its natural frequencies; compression will reduce these quantities.

Example 32

Find the first three natural frequencies of a steel bar 3 cm in diameter, which is simply supported at each end, and has a length of 1.5 m. Take $\rho = 7780$ kg/m^3 and $E = 208$ GN/m^2.

For the bar,

$$\sqrt{\left(\frac{EI}{A\rho}\right)} = \sqrt{\left(\frac{208 \times 10^9 \times \pi(0.03)^4/64}{\pi(0.03/2)^2\ 7780}\right)}\ \text{m/s}^2 = 38.8\ \text{m/s}^2.$$

Thus

$$\omega_1 = \frac{\pi^2}{1.5^2}\ 38.8 = 170.2\ \text{rad/s}\quad \text{and}\quad f_1 = 27.1\ \text{Hz}.$$

Hence

$$f_2 = 27.1 \times 4 = 108.4\ \text{Hz}$$

and

$$f_3 = 27.1 \times 9 = 243.8\ \text{Hz}.$$

If the beam is subjected to an axial tension T, the modified equation of motion leads to the following expression for the natural frequencies:

$$\omega_n^2 = \left(\frac{n\pi}{l}\right)^2 \frac{T}{A\rho} + \left(\frac{n\pi}{l}\right)^4 \frac{EI}{A\rho}.$$

For the case when $T = 1000$ N the correction to ω_n^2 is ω_c^2, where

$$\omega_c^2 = \left(\frac{\pi}{1.5}\right)^2 \left(\frac{1000}{\pi(0.03/2)^2\ 7780}\right) = 795\ (\text{rad/s})^2.$$

That is, $f_c = 4.5$ Hz. Hence $f_1 = \sqrt{(4.5^2 + 27.1^2)} = 27.5$ Hz.

4.2.4 Transverse vibration of a beam with discrete bodies

In those cases where it is required to find the lowest frequency of transverse vibration of a beam that carries discrete bodies, Dunkerley's method may be used. This is a simple analytical technique which enables a wide range of vibration problems to be solved using a hand calculator. Dunkerley's method uses the following equation:

$$\frac{1}{\omega_1^2} \simeq \frac{1}{P_1^2} + \frac{1}{P_2^2} + \frac{1}{P_3^2} + \frac{1}{P_4^2} + ...,$$

where ω_1 is the lowest natural frequency of a system and P_1, P_2, P_3, ... are the frequencies of each body acting alone (see section 3.2.1.2).

Example 33

A steel shaft ($\rho = $ 8000 kg/m³, $E = $ 210 GN/m²) 0.055 m diameter, running in self-aligning bearings 1.25 m apart, carries a rotor of mass 70 kg, 0.4 m from one bearing. Estimate the lowest critical speed.

For the shaft alone

$$\sqrt{\left(\frac{EI}{A\rho}\right)} = \sqrt{\left(\frac{210 \times 10^9 \times \pi(0.055)^4/64}{\pi(0.055/2)^2 \ 8000}\right)} = 70.45 \ \text{m/s}^2.$$

Thus $P_1 = \left(\dfrac{\pi}{1.25}\right)^2 70.45 = 445 \ \text{rad/s} = 4249 \ \text{rev/min}.$

This is the lowest critical speed for the shaft without the rotor. For the rotor alone, neglecting the mass of the shaft,

$$P_2 = \sqrt{(k/m)} \ \text{rad/s}$$

and

$$k = 3EIl/(x^2(l-x)^2),$$

where $x = $ 0.4 m and $l = $ 1.25 m.
Thus

$$k = 3.06 \ \text{MN/m}$$

and

$$P_2 = \sqrt{((3.06 \times 10^6)/70)}$$

$$= 209.1 \ \text{rad/s} = 1996 \ \text{rev/min}.$$

Now using Dunkerley's method,

$$1/N_1^2 = 1/4249^2 + 1/1996^2, \text{ hence } N_1 = 1807 \ \text{rev/min}.$$

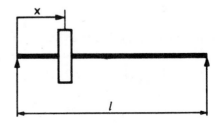

4.2.5 Receptance analysis

Many structures can be considered to consist of a number of beams fastened together. Thus if the receptances of each beam are known, the frequency equation of the structure can easily be found by carrying out a subsystem analysis (section 3.2.3). The required receptances can be found by inserting the appropriate boundary conditions in the general solution to the beam equation.

It will be appreciated that this method of analysis is ideal for computer solutions because of its repetitive nature.

For example, consider a beam that is pinned at one end ($x = 0$) and free at the other end ($x = l$). This type of beam is not commonly used in practice, but it is useful for analysis purposes. With a harmonic moment of amplitude M applied to the pinned end,

at $x = 0$, $X = 0$ (zero deflection) and

$$\frac{d^2X}{dx^2} = \frac{M}{EI} \text{ (bending moment } M\text{)},$$

and at $x = l$,

$$\frac{d^2X}{dx^2} = 0 \text{ (zero bending moment)}$$

and

$$\frac{d^3X}{dx^3} = 0 \text{ (zero shear force)}.$$

Now, in general,

$$X = C_1 \cos \lambda x + C_2 \sin \lambda x + C_3 \cosh \lambda x + C_4 \sinh \lambda x.$$

Thus applying these boundary conditions,

$$0 = C_1 + C_3 \quad \text{and} \quad \frac{M}{EI} = -C_1\lambda^2 + C_3\lambda^2.$$

Also

$$0 = -C_1\lambda^2 \cos \lambda l - C_2\lambda^2 \sin \lambda l + C_3\lambda^2 \cosh \lambda l + C_4\lambda^2 \sinh \lambda l.$$

and

$$0 = C_1\lambda^3 \sin \lambda l - C_2\lambda^3 \cos \lambda l + C_3\lambda^3 \sinh \lambda l + C_4\lambda^3 \cosh \lambda l.$$

By solving these four equations $C_{1,2,3,4}$ can be found and substituted into the general solution. It is found that the receptance moment/slope at the pinned end is

$$\frac{(1 + \cos \lambda l \cosh \lambda l)}{EI\lambda (\cos \lambda l \sinh \lambda l - \sin \lambda l \cosh \lambda l)}$$

and at the free end is

$$\frac{2 \cos \lambda l \cosh \lambda l}{EI\lambda (\cos \lambda l \sinh \lambda l - \sin \lambda l \cosh \lambda l)}.$$

The frequency equation is given by

$$\cos \lambda l \sinh \lambda l - \sin \lambda l \cosh \lambda l = 0,$$

that is, $\tan \lambda l = \tanh \lambda l$.

Moment/deflection receptances can also be found.

By inserting the appropriate boundary conditions into the general solution, the receptance due to a harmonic moment applied at the free end, and harmonic forces applied to either end, can be deduced. Receptances for beams with all end conditions are tabulated in *The Mechanics of Vibration* by R. E. D. Bishop & D. C. Johnson (CUP, 1960/79), thereby greatly increasing the ease of applying this technique.

Example 34

A hinged beam structure is modelled by the array shown below:

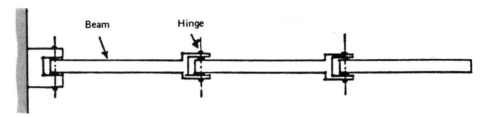

The hinges are pivots with torsional stiffness k_T and their mass is negligible. All hinges and beams are the same.

It is required to find the natural frequencies of free vibration of the array, so that the excitation of these frequencies, and therefore resonance, can be avoided.

Since all the beams are identical, the receptance technique is relevant for finding the frequency equation. This is because the receptances of each subsystem are the same, which leads to some simplification in the analysis.

There are two approaches:

(i)   to split the array into subsystems comprising torsional springs and beams,
(ii)   to split the array into subsystems comprising spring–beam assemblies.

This approach results in a smaller number of subsystems.

Considering the first approach, and only the first element of the array, the subsystems could be either

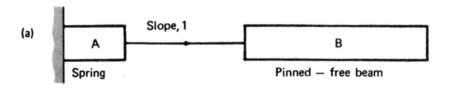

or

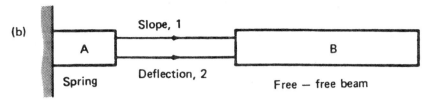

For (a) the frequency equation is $\alpha_{11} + \beta_{11} = 0$, whereas for (b) the frequency equation is

$$\begin{vmatrix} \alpha_{11} + \beta_{11} & \alpha_{12} + \beta_{12} \\ \alpha_{21} + \beta_{21} & \alpha_{22} + \beta_{22} \end{vmatrix} = 0,$$

where α_{11} is the moment/slope receptance for A, β_{11} is the moment/slope receptance for B, β_{12} is the moment/deflection receptance for B, β_{22} is the force/deflection receptance for B, and so on.

For (a), either calculating the beam receptances as above, or obtaining them from tables, the frequency equation is

$$\frac{1}{k_T} + \frac{\cos \lambda l \cosh \lambda l + 1}{EI\lambda(\cos \lambda l \sinh \lambda l - \sin \lambda l \cosh \lambda l)} = 0,$$

where

$$\lambda = \sqrt[4]{\left(\frac{A\rho\omega^2}{EI}\right)}.$$

For (b), the frequency equation is

$$\begin{vmatrix} \dfrac{1}{k_T} + \dfrac{\cos \lambda l \sinh \lambda l + \sin \lambda l \cosh \lambda l}{EI\lambda(\cos \lambda l \cosh \lambda l - 1)} & \dfrac{-\sin \lambda l \sinh \lambda l}{EI\lambda^2(\cos \lambda l \cosh \lambda l - 1)} \\[3mm] \dfrac{-\sin \lambda l \sinh \lambda l}{EI\lambda^2(\cos \lambda l \cosh \lambda l - 1)} & \dfrac{-(\cos \lambda l \sinh \lambda l - \sin \lambda l \cosh \lambda l)}{EI\lambda^3(\cos \lambda l \cosh \lambda l - 1)} \end{vmatrix} = 0,$$

which reduces to the equation given by method (a).

The frequency equation has to be solved after inserting the structural parameters, to yield the natural frequencies of the structure.

For the whole array it is preferable to use approach (ii), because this results in a smaller number of subsystems than (i), with a consequent simplification of the frequency equation. However, it will be necessary to calculate the receptances of the spring pinned–free beam if approach (ii) is adopted.

The analysis of structures such as frameworks can also be accomplished by the receptance technique, by dividing the framework to be analysed into beam substructures. For example, if the in-plane natural frequencies of a portal frame are required, it can be

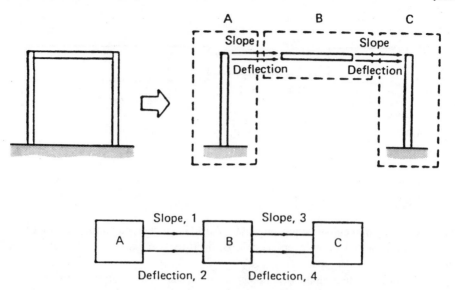

Fig. 4.4. Portal frame substructure analysis.

divided into three substructures coupled by the conditions of compatibility and equilibrium, as shown in Fig. 4.4.

Substructures A and C are cantilever beams undergoing transverse vibration, whereas B is a free–free beam undergoing transverse vibration. Beam B is assumed rigid in the horizontal direction, and the longitudinal deflection of beams A and C is assumed to be negligible.

Because the horizontal member B has no coupling between its horizontal and flexural motion $\beta_{12} = \beta_{14} = \beta_{23} = \beta_{34} = 0$, so that the frequency equation becomes

$$\begin{vmatrix} \alpha_{11} + \beta_{11} & \alpha_{11} & \beta_{13} & 0 \\ \alpha_{21} & \alpha_{22} + \beta_{22} & 0 & \beta_{24} \\ \beta_{31} & 0 & \gamma_{33} + \beta_{33} & \beta_{34} \\ 0 & \beta_{42} & \gamma_{43} & \gamma_{44} + \beta_{44} \end{vmatrix} = 0.$$

4.3 THE ANALYSIS OF CONTINUOUS STRUCTURES BY RAYLEIGH'S ENERGY METHOD

Rayleigh's method, as described in section 2.1.4, gives the lowest natural frequency of transverse beam vibration as

$$\omega^2 = \frac{\displaystyle\int EI\left(\frac{d^2y}{dx^2}\right)^2 dx}{\displaystyle\int y^2 \, dm}.$$

A function of x representing y can be determined from the static deflected shape of the beam, or a suitable part sinusoid can be assumed, as shown in the following examples.

Example 35

A simply supported beam of length l and mass m_2 carries a body of mass m_1 at its mid-point. Find the lowest natural frequency of transverse vibration.

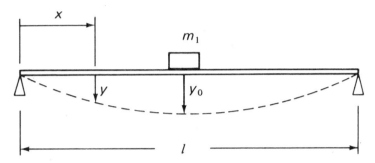

This example has been fully discussed above (Example 4, p. 25). However, the Dunkerley method can also be used. Here

$$P_1^2 = \frac{48\ EI}{m_1 l^3} \quad \text{and} \quad P_2^2 = \frac{EI\ \pi^4}{m_2 l^3}.$$

Thus

$$\frac{1}{\omega^2} = \frac{m_1 l^3}{48\ EI} + \frac{m_2 l^3}{\pi^4\ EI}.$$

Hence

$$\omega^2 = \frac{EI\left(\dfrac{\pi}{l}\right)^4 \dfrac{l}{2}}{\left(1.015 m_1 + \dfrac{m_2}{2}\right)},$$

which is very close to the value determined by the Rayleigh method.

Example 36

A pin-ended strut of length l has a vertical axial load P applied. Determine the frequency of free transverse vibration of the strut, and the maximum value of P for stability. The strut has a mass m and a second moment of area I, and is made from material with modulus of elasticity E.

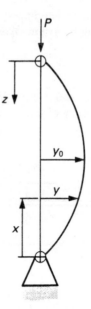

The deflected shape can be expressed by

$$y = y_0 \sin \pi \frac{x}{l},$$

since this function satisfies the boundary conditions of zero deflection and bending moment at $x = 0$ and $x = l$.

Now,

$$V_{max} = \frac{1}{2} \int EI \left(\frac{d^2 y}{dx^2} \right)^2 dx - Pz,$$

where

$$\frac{1}{2} \int EI \left(\frac{d^2 y}{dx^2} \right)^2 dx = \frac{1}{2} \int_0^l EI \left(\frac{\pi}{l} \right)^4 y_0^2 \sin^2 \pi \frac{x}{l} dx$$

$$= \frac{EI}{4} \frac{\pi^4}{l^3} y_0^2,$$

and

$$z = \int_0^l \left(\sqrt{\left(1 + \left(\frac{dy}{dx} \right)^2 \right)} - 1 \right) dx$$

$$= \int_0^l \frac{1}{2} \left(\frac{dy}{dx} \right)^2 dx$$

$$= \frac{1}{2} \int_0^l y_0^2 \left(\frac{\pi}{l} \right)^2 \cos^2 \pi \frac{x}{l} \, dx$$

$$= \frac{y_0^2}{4} \frac{\pi^2}{l}.$$

Thus

$$V_{max} = \left(\frac{EI}{4} \frac{\pi^4}{l^3} - \frac{P}{4} \frac{\pi^2}{l} \right) y_0^2.$$

Now,

$$T_{max} = \frac{1}{2} \int y^2 \, dm = \frac{1}{2} \int_0^l y^2 \frac{m}{l} \, dx$$

$$= \frac{1}{2} \int_0^l y_0^2 \sin^2 \pi \frac{x}{l} \frac{m}{l} \, dx = \frac{m}{4} y_0^2.$$

Thus

$$\omega^2 = \frac{\left(\dfrac{EI}{4} \dfrac{\pi^4}{l^3} - \dfrac{P}{4} \dfrac{\pi^2}{l} \right)}{\dfrac{m}{4}},$$

and

$$f = \frac{1}{2} \sqrt{ \left(\frac{EI \, (\pi/l)^2 - P}{ml} \right) } \text{ Hz.}$$

From section 2.1.4, for stability

$$\frac{dV}{dy_0} = 0 \quad \text{and} \quad \frac{d^2V}{dy_0^2} > 0,$$

that is,

$$y_0 = 0 \text{ and } EI \frac{\pi^2}{l^2} > P;$$

$y_0 = 0$ is the equilibrium position about which vibration occurs, and $P < EI\,\pi^2/l^2$ is the necessary condition for stability. $EI\,\pi^2/l^2$ is known as the Euler buckling load.

4.4   TRANSVERSE VIBRATION OF THIN UNIFORM PLATES

Plates are frequently used as structural elements so that it is sometimes necessary to analyse plate vibration. The analysis considered will be restricted to the vibration of thin uniform flat plates. Non-uniform plates that occur in structures, for example, those which are ribbed or bent, may best be analysed by the finite element technique, although exact theory does exist for certain curved plates and shells.

The analysis of plate vibration represents a distinct increase in the complexity of vibration analysis, because it is necessary to consider vibration in two dimensions instead of the single-dimension analysis carried out hitherto. It is essentially therefore, an introduction to the analysis of the vibration of multi-dimensional structures.

Consider a thin uniform plate of an elastic, homogeneous isotropic material of thickness h, as shown in Fig. 4.5.

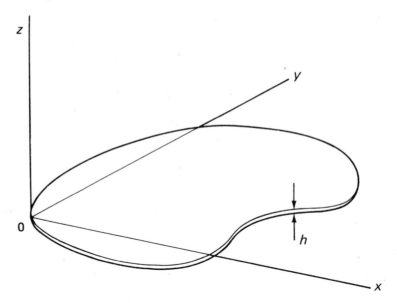

Fig. 4.5.   Thin uniform plate.

If v is the deflection of the plate at a point (x, y), then it is shown in *Vibration Problems in Engineering* by S. Timoshenko (Van Nostrand, 1974), that the potential energy of bending of the plate is

$$\frac{D}{2}\iint\left\{\left(\frac{\partial^2 v}{\partial x^2}\right)^2 + \left(\frac{\partial^2 v}{\partial y^2}\right)^2 + 2v\,\frac{\partial^2 v}{\partial x^2}\,\frac{\partial^2 v}{\partial y^2} + 2\,(1-v)\left(\frac{\partial^2 v}{\partial x.\partial y}\right)^2\right\}\,dx\,dy$$

where the flexural rigidity,

$$D = \frac{Eh^3}{12(1 - v^2)}$$

and v is Poisson's Ratio.

The kinetic energy of the vibrating plate is

$$\frac{\rho h}{2} \iint \dot{v}^2 \, dx \, dy,$$

where ρh is the mass per unit area of the plate.

In the case of a rectangular plate with sides of length a and b, and with simply supported edges, at a natural frequency ω, v can be represented by

$$v = \phi \sin m\pi \frac{x}{a} \sin n\pi \frac{y}{b},$$

where ϕ is a function of time.

Thus

$$V = \frac{\pi^4 ab}{8} D \phi^2 \left(\frac{m^2}{a^2} + \frac{n^2}{b^2} \right)^2$$

and

$$T = \frac{\rho h}{2} \frac{ab}{4} \dot{\phi}^2.$$

Since $d(T + V)/dt = 0$ in a conservative structure,

$$\frac{\rho h}{2} \frac{ab}{4} 2\dot{\phi}\ddot{\phi} + \frac{\pi^4 ab}{8} D \, 2\phi\dot{\phi} \left(\frac{m^2}{a^2} + \frac{n^2}{b^2} \right)^2 = 0;$$

that is, the equation of motion is

$$\rho h \ddot{\phi} + \pi^4 D \left(\frac{m^2}{a^2} + \frac{n^2}{b^2} \right)^2 \phi = 0.$$

Thus ϕ represents simple harmonic motion and

$$\phi = A \sin \omega_{mn} t + B \cos \omega_{mn} t,$$

where

$$\omega_{mn} = \pi^2 \sqrt{\left(\frac{D}{\rho h} \right)} \left[\frac{m^2}{a^2} + \frac{n^2}{b^2} \right] \text{ rad/s.}$$

Now,

$$v = \phi \sin m\pi \frac{x}{a} \sin n\pi \frac{y}{b},$$

thus $v = 0$ when $\sin m\pi x/a = 0$ or $\sin n\pi y/b = 0$, and hence the plate has nodal lines when vibrating in its normal modes.

Typical nodal lines of the first six modes of vibration of a rectangular plate, simply supported on all edges, are shown in Fig. 4.6.

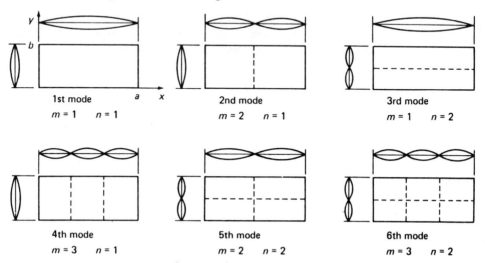

Fig. 4.6.   Transverse plate vibration mode shapes.

An exact solution is only possible using this method if two opposite edges of the plate are simply supported: the other two edges can be free, hinged or clamped. If this is not the case, for example if the plate has all edges clamped, a series solution for v has to be adopted.

For a simply supported square plate of side $a(=b)$, the frequency of free vibration becomes

$$f = \pi \frac{m^2}{a^2} \sqrt{\left(\frac{D}{\rho h}\right)} \text{ Hz,}$$

whereas for a square plate simply supported along two opposite edges and free on the others,

$$f = \frac{\alpha}{2\pi a^2} \sqrt{\left(\frac{D}{\rho h}\right)} \text{ Hz,}$$

where $\alpha = 9.63$ in the first mode (1, 1), $\alpha = 16.1$ in the second mode (1, 2), and $\alpha = 36.7$ in the third mode (1, 3).

Thus the lowest, or fundamental, natural frequency of a simply supported/free square plate of side l and thickness d is

$$\frac{9.63}{2\pi l^2}\sqrt{\left(\frac{Ed^3}{12(1-v^2)\,\rho d}\right)} = \frac{10.09}{2\pi l^2}\sqrt{\left(\frac{Ed^2}{12\rho}\right)}\mathrm{Hz},$$

if $v = 0.3$.

The theory for beam vibration gives the fundamental natural frequency of a beam simply supported at each end as

$$\frac{1}{2\pi}\left(\frac{\pi}{l}\right)^2\sqrt{\left(\frac{EI}{A\rho}\right)}\mathrm{Hz}.$$

If the beam has a rectangular section $b \times d$, $I = \dfrac{bd^3}{12}$ and $A = bd$.

Thus

$$f = \frac{1}{2\pi}\left(\frac{\pi}{l}\right)^2\sqrt{\left(\frac{Ed^2}{12\rho}\right)}\mathrm{Hz},$$

that is,

$$f = \frac{9.86}{2\pi l^2}\sqrt{\left(\frac{Ed^2}{12\rho}\right)}\mathrm{Hz}.$$

This is very close (within about 2%) to the frequency predicted by the plate theory, although of course beam theory cannot be used to predict all the higher modes of plate vibration, because it assumes that the beam cross section is not distorted. Beam theory becomes more accurate as the aspect ratio of the beam, or plate, increases.

For a circular plate of radius a, clamped at its boundary, it has been shown that the natural frequencies of free vibration are given by

$$f = \frac{\alpha}{2\pi a^2}\sqrt{\left(\frac{D}{\rho h}\right)}\mathrm{Hz},$$

where α is as given in Table 4.2.

Table 4.2

Number of nodal circles	Number of nodal diameters		
	0	1	2
0	10.21	21.26	34.88
1	39.77	60.82	84.58
2	89.1	120.08	153.81
3	158.18	199.06	242.71

The vibration of a wide range of plate shapes with various types of support is fully discussed in NASA publication SP-160 *Vibration of Plates* by A. W. Leissa.

4.5   THE FINITE ELEMENT METHOD

Many structures, such as a ship hull or engine crankcase, are too complicated to be analysed by classical techniques, so that an approximate method has to be used. It can be seen from the receptance analysis of complicated structures that breaking a dynamic structure down into a large number of substructures is a useful analytical technique, provided that sufficient computational facilities are available to solve the resulting equations. The finite element method of analysis extends this method to the consideration of continuous structures as a number of elements, connected to each other by conditions of compatibility and equilibrium. Complicated structures can thus be modelled as the aggregate of simpler structures.

The principal advantage of the finite element method is its generality; it can be used to calculate the natural frequencies and mode shapes of any linear elastic system. However, it is a numerical technique that requires a fairly large computer, and care has to be taken over the sensitivity of the computer output to small changes in input.

For beam type systems the finite element method is similar to the lumped mass method, because the system is considered to be a number of rigid mass elements of finite size connected by massless springs. The infinite number of degrees of freedom associated with a continuous system can thereby be reduced to a finite number of degrees of freedom, which can be examined individually.

The finite element method therefore consists of dividing the dynamic system into a series of elements by imaginary lines, and connecting the elements only at the intersections of these lines. These intersections are called nodes. It is unfortunate that the word node has been widely accepted for these intersections; this meaning should not be confused with the zero vibration regions referred to in vibration analysis. The stresses and strains in each element are then defined in terms of the displacements and forces at the nodes, and the mass of the elements is lumped at the nodes. A series of equations is thus produced for the displacement of the nodes and hence the system. By solving these equations the stresses, strains, natural frequencies and mode shapes of the system can be determined. The accuracy of the finite element method is greatest in the lower modes, and increases as the number of elements in the model increases. The finite element method of analysis is considered in *The Finite Element Method* by O. C. Zienkiewicz (McGraw Hill, 1977) and *A First Course in Finite Element Analysis* by Y. C. Pao (Allyn and Bacon, 1986).

4.6   THE VIBRATION OF BEAMS FABRICATED FROM MORE THAN ONE MATERIAL

Engineering structures are sometimes fabricated using composite materials. These applications are usually where high strength and low weight are required as, for example, in aircraft, space vehicles and racing cars. Composite materials are produced by embedding high-strength fibres in the form of filaments or yarn in a plastic, metal or ceramic matrix. They are more expensive than conventional materials but their application or manufacturing methods often justify their use.

The most common plastic materials used are polyester and epoxy resin, reinforced with glass. The glass may take the form of strands, fibres or woven fabrics. The desirable quality of glass fibres is their high tensile strength. Naturally the orientation and alignment or otherwise of the fibres can greatly affect the properties of the composite. Glass reinforced plastic (GRP) is used in such structures as boats, footbridges and car bodies. Boron fibres are more expensive than glass but because they are six times stiffer they are sometimes used in critical applications.

Carbon fibres are expensive, but they combine increased stiffness with a very high tensile strength, so that composites of carbon fibre and resin can have the same tensile strength as steel but weigh only a quarter as much. Because of this carbon fibre composites now compete directly with aluminium in many aircraft structural applications. Cost precludes its large-scale use, but in the case of the A320 Airbus, for example, over 850 kg of total weight is saved by using composite materials for control surfaces such as flaps, rudder, fin and elevators in addition to some fairings and structural parts.

Analysis of the vibration of such structural components can be conveniently carried out by the finite element method (section 4.5), or more usefully by the modal analysis method (section 3.3). However, composite materials are usually anisotropic so the analysis can be difficult. Inherent damping is often high however, even though it may be hard to predict due to variations in such factors as manufacturing techniques and fibre/matrix wetting.

Concrete is usually reinforced by steel rods, bars or mesh to contribute tensile strength. In reinforced concrete, the tensile strength of the steel supplements the compressive strength of the concrete to provide a structural member capable of withstanding high stresses of all kinds over large spans. It is a fairly cheap material and is widely used in the construction of bridges, buildings, boats, structural frameworks and roads.

It is sometimes appropriate, therefore, to fabricate structural components such as beams, plates and shells from more than one material, either in whole or in part, to take advantage of the different and supplementary properties of the two materials. Composites are also sometimes incorporated into highly stressed parts of a structure by applying patches of a composite to critical areas.

The vibration analysis of composite structures can be lengthy and difficult, but the fundamental frequency of vibration of a beam made from two materials can be determined using the energy principle, as follows.

Fig. 4.7 shows a cross section through a beam made from two materials 1 and 2 bonded at a common interface. Provided the bond is sufficiently good to prevent relative slip, a plane section before bending remains plane after bending so that the strain distribution is linear across the section, although the normal stress will change at the interface because of the difference in the elastic moduli of the two materials E_1 and E_2.

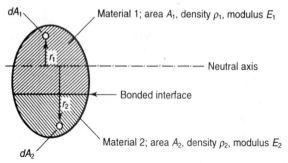

Fig. 4.7. Composite beam cross section.

Now, from section 2.1.4.2 and Fig. 2.11, the strain ε at a distance r from the neutral axis of a beam in bending is

$$\frac{(R + r)d\theta - Rd\theta}{Rd\theta} = \frac{r}{R}.$$

Hence the strain at a distance r_1 from the neutral axis is

$$\varepsilon_1 = r_1 \frac{d^2y}{dx^2},$$

and similarly

$$\varepsilon_2 = r_2 \frac{d^2y}{dx^2}.$$

Hence, the corresponding stresses are

$$\sigma_1 = E_1\varepsilon_1 = E_1r_1 \frac{d^2y}{dx^2}$$

and

$$\sigma_2 = E_2\varepsilon_2 = E_2r_2 \frac{d^2y}{dx^2}.$$

The strain energy stored in the two materials per unit volume is $dV_1 + dV_2$ where

$$dV_1 = \frac{\sigma_1 \varepsilon_1}{2} = \frac{E_1}{2} r_1^2 \left(\frac{d^2 y}{dx^2}\right)^2$$

and

$$dV_2 = \frac{-\sigma_2 \varepsilon_2}{2} = \frac{E_2}{2} r_2^2 \left(\frac{d^2 y}{dx^2}\right)^2.$$

Integrating over the volume of a beam of length l gives

$$V_{max} = \frac{E_1}{2} \int_0^l \int r_1^2 \left(\frac{d^2 y}{dx^2}\right)^2 dA_1 \, dx + \frac{E_2}{2} \int_0^l \int r_2^2 \left(\frac{d^2 y}{dx^2}\right)^2 dA_2 \, dx.$$

Now

$$I_1 = \int r_1^2 \, dA_1 \text{ and } I_2 = \int r_2^2 \, dA_2,$$

so

$$V_{max} = \left(\frac{E_1 I_1 + E_2 I_2}{2}\right) \int_0^l \left(\frac{d^2 y}{dx^2}\right)^2 dx.$$

I_1 and I_2 can only be calculated when the location of the neutral axis of the composite cross section is known. This can be found using an equivalent cross section for one material.

The mass per unit length of the composite is $\rho_1 A_1 + \rho_2 A_2$, so that

$$T_{max} = \left(\frac{\rho_1 A_1 + \rho_2 A_2}{2}\right) \int_0^l (y\omega)^2 \, dx.$$

A shape function has therefore to be assumed before T_{max} can be calculated.

Putting $T_{max} = V_{max}$ gives the natural frequency ω.

Example 37

A simply supported beam of length l is fabricated from two materials M1 and M2. Find the fundamental natural frequency of the beam using Rayleigh's method and the shape function

$$y = P \sin\left(\frac{\pi x}{l}\right).$$

$$V_{max} = \left(\frac{E_{M1} I_{M1} + E_{M2} I_{M2}}{2}\right) \int_0^l \left(\frac{d^2 y}{dx^2}\right)^2 dx$$

$$= \left(\frac{E_{M1} I_{M1} + E_{M2} I_{M2}}{2}\right) \frac{P^2 \pi^4}{2l^3}$$

$$T_{max} = \left(\frac{\rho_{M1} A_{M1} + \rho_{M2} A_{M2}}{2} \right) \int_0^l (y\omega)^2 \, dx$$

$$= \left(\frac{\rho_{M1} A_{M1} + \rho_{M2} A_{M2}}{2} \right) \frac{P^2 l}{2} \omega^2.$$

Putting $T_{max} = V_{max}$ gives

$$\omega^2 = \left(\frac{E_{M1} I_{M1} + E_{M2} I_{M2}}{\rho_{M1} A_{M1} + \rho_{M2} A_{M2}} \right) \frac{\pi^4}{l^4}$$

So that

$$\omega = \frac{\pi^2}{l^2} \sqrt{\left(\frac{E_{M1} I_{M1} + E_{M2} I_{M2}}{\rho_{M1} A_{M1} + \rho_{M2} I_{M2}} \right)} \text{ rad/s.}$$

I_{M1} and I_{M2} can be calculated once the position of the neutral axis has been found.

This method of analysis can obviously be extended to beams fabricated from more than two materials.

5

Damping in structures

5.1 SOURCES OF VIBRATION EXCITATION AND ISOLATION

Before attempting to reduce the vibration levels in a machine or structure by increasing its damping, every effort should be made to reduce the vibration excitation at its source. It has to be accepted that many machines and processes generate a disturbing force of one sort or another, but the frequency of the disturbing force should not be at, or near, a natural frequency of the structure otherwise resonance will occur, with the resulting high amplitudes of vibration and dynamic stresses, and noise and fatigue problems. Resonance may also prevent the structure fulfilling the desired function.

Some reduction in excitation can often be achieved by changing the machinery generating the vibration, but this can usually only be done at the design stage. Re-siting equipment may also effect some improvement. However, structural vibration caused by external excitation sources such as ground vibration, cross winds or turbulence from adjacent buildings can only be controlled by damping.

In some machines vibrations are deliberately excited as part of the process, for example, in vibratory conveyors and compactors, and in ultrasonic welding. Naturally, nearby machines have to be protected from these vibrations.

Rotating machinery such as fans, turbines, motors and propellers can generate disturbing forces at several different frequencies such as the rotating speed and blade passing frequency. Reciprocating machinery such as compressors and engines can rarely be perfectly balanced, and an exciting force is produced at the rotating speed and at harmonics. Strong vibration excitation in structures can also be caused by pressure fluctuation in gases and liquids flowing in pipes, as well as intermittent loads such as those imposed by lifts in buildings.

There are two basic types of structural vibration: steady-state vibration caused by continually running machines such as engines, air-conditioning plants and generators either within the structure or situated in a neighbouring structure, and transient vibration caused by a short-duration disturbance such as a lorry or train passing over an expansion joint in a road or over a bridge.

Some relief from steady-state vibration excitation can often be gained by moving the source of the excitation, since the mass of the vibration generator has some effect on the natural frequencies of the supporting structure. For example, in a building it may be an advantage to move mechanical equipment to a lower floor, and in a ship re-siting propulsion or service machinery may prove effective. The effect of local stiffening of the structure may prove to be disappointing, however, because by increasing the stiffness the mass is also increased, so that the change in the $\sqrt{(k/m)}$ may prove to be very small.

Occasionally a change in the vibration generating equipment can reduce vibration levels. For example a change in gear ratios in a mechanical drive system, or a change from a four-bladed to a three-bladed propeller in a ship propulsion system will alter the excitation frequency provided the speed of rotation is not changed. However, in many cases the running speeds of motors and engines are closely controlled as in electric generator sets, so there is no opportunity for changing the excitation frequency.

If vibration excitation cannot be reduced to acceptable levels, so that the system response is still too large, some measure of vibration isolation may be necessary (see section 2.3.2.1).

5.2   VIBRATION ISOLATION

It is shown in section 2.3.2.1 that good vibration isolation, that is, low force and motion transmissibility, can be achieved by supporting the vibration generator on a flexible low-frequency mounting. Thus although disturbing forces are generated, only a small proportion of them are transmitted to the supporting structure. However, this theory assumes that a mode of vibration is excited by a harmonic force passing through the centre of mass of the installation; although this is often a reasonable approximation it rarely actually occurs in practice because, due to a lack of symmetry of the supported machine, several different mountings may be needed to achieve a level installation, and the mass centre is seldom in the same plane as the tops of the mountings.

Thus the mounting which provides good isolation against a vertical exciting force may allow excessive horizontal motion, because of a frequency component close to the natural frequency of the horizontal mode of vibration. Also a secondary exciting force acting eccentrically from the centre of mass can excite large rotation amplitudes when the frequency is near to that of a rocking mode of an installation.

To limit the motion of a machine installation that generates harmonic forces and moments, the mass and inertia of the installation supported by the mountings may have to be increased; that is, an inertia block may have to be added to the installation. If non-metallic mountings are used the dynamic stiffness at the frequencies of interest will have to be found, probably by carrying out further dynamic tests in which the mounting is correctly loaded; they may also possess curious damping characteristics which may be included in the analysis by using the concept of complex stiffness, as discussed in section 2.2.5.

Air bags or bellows are sometimes used for very low-frequency mountings where some swaying of the supported system is acceptable. This is an important consideration because if the motion of the inertia block and the machinery is large, pipework and other services may be overstressed, which can lead to fatigue failure of these components. Approximate analysis shows that the natural frequency of a body supported on bellows filled with air under pressure is inversely proportional to the square root of the volume of the bellows, so that a change in natural frequency can be effected simply by a change in bellows volume. This can easily be achieved by opening or closing valves connecting the bellows to additional receivers, or by adding a liquid to the bellows. Natural frequencies of 0.5 Hz, or even less, are obtainable. An additional advantage of air suspension is that the system can be made self-levelling, when fitted with suitable valves and an air supply. Air pressures of about 5–10 times atmospheric pressure are usual.

Greater attenuation of the exciting force at high frequencies can be achieved by using a two-stage mounting. In this arrangement the machine is set on flexible mountings on an inertia block, which is itself supported by flexible mountings. This may not be too expensive to install since in many cases an existing subframe or structure can be used as the inertia block. If a floating floor in a building is used as the inertia block, some allowance must be made for the additional stiffness arising from the air space below it. This can be found by measuring the dynamic stiffness of the floor by means of resonance tests.

Naturally, techniques used for isolating structures from exciting forces arising in machinery and plant can also be used for isolating delicate equipment from vibrations in the structure. For example, sensitive electrical equipment in ships can be isolated from hull vibration, and operating tables and metrology equipment can be isolated from building vibration.

The above isolation systems are all *passive*; an *active* isolation system is one in which the exciting force or moment is applied by an externally powered force or couple. The opposing force or moment can be produced by means such as hydraulic rams, out-of-balance rotating bodies or electromagnetism. Naturally it is essential to have accurate phase and amplitude control, to ensure that the opposing force is always equal, and opposite, to the exciting force. Although active isolation systems can be expensive to install, excellent results are obtainable so that the supporting structure is kept almost completely still. However it must be noted that force actuators such as hydraulic rams must react on another part of the system.

If, after careful selection and design of machinery and equipment, careful installation and commissioning and carrying out isolation as necessary, the vibration levels in the system are still too large, then some increase in the damping is necessary. This is also the case when excitation occurs from sources beyond the designer's control such as cross winds, earthquakes and currents.

5.3   STRUCTURAL VIBRATION LIMITS

The vibration to which a structure may be subjected is usually considered with respect to its effect on the structure itself, and not on its occupants, equipment or machinery. Modern structures are less massive and have lower damping than hitherto, and because of the sophisticated design and analysis techniques now used they generally have less re-dundancy. The consideration of vibration limits is therefore becoming increasingly important for the maintenance of structural integrity and fitness for purpose. It is important to appreciate that even when the level of structural vibration is considered intolerable by the occupants, the risk of structural damage from sustained vibration is usually very small. In some cases vibration limits may have to be set in accordance with operator or occupant criteria which will be well below those which would cause structural damage, as discussed later.

Structural vibration limits for particular damage risks can be classified according to the level of vibration intensity, or by consideration of the largest of the rms or peak velocities measured in one of three orthogonal directions.

5.3.1 Vibration intensity

Fig. 5.1 shows limit lines in terms of vibration amplitude and frequency for various levels of damage. These lines correspond to constant values of $X^2 f^3$ where X is the amplitude of harmonic vibration and f is the frequency. For harmonic motion $\ddot{X} = Xf^2(2\pi)^2$ so that the vibration intensity Z is

$$Z = \frac{\ddot{X}^2}{f} = 16\pi^4 X^2 f^3.$$

If the reference value for Z, Z_0 is taken to be 10 mm^2/s^3 the dimensionless vibration

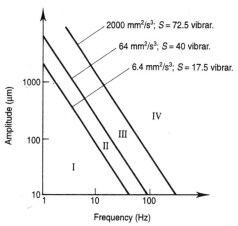

Region
I – No damage
II – Possible plaster cracks
III – Probable damage to load-bearing structural parts
IV – Damage to load-bearing parts
IV+ – Destruction

Fig. 5.1.   Structural damage limits.

intensity S is given by

$$S = 10 \log \frac{Z}{Z_0} = 22 \log (X^2 f^3) \text{ vibrar.}$$

There seems to be little risk of structural damage for values of $X^2 f^3$ below 50 mm^2/s^3 ($S = 37.4$ vibrar).

The allowable limit for building vibrations is usually taken to lie in the range 30–40 vibrar which corresponds to an rms velocity of about 5 mm/s at frequencies between 5 and 50 Hz. It has been found that rms velocity is probably a more realistic criterion for damage to present day structures than vibration intensity.

5.3.2 Vibration velocity

Fig. 5.2, which expresses harmonic vibration amplitude as a function of frequency shows that lines for constant velocity have smaller slopes than lines for constant vibration intensity. Therefore standards based on constant velocity give increased weight to lower-frequency vibrations which are more likely to induce structural resonance and damage than frequencies above 50 Hz. An rms based quantity provides a reliable criterion for damage evaluation since it is related to vibrational energy levels.

Conventional types of structure do not usually experience any damage from steady-state vibration with a peak velocity (V_P) less than 10 mm/s, ($V_{rms} = 7$ mm/s). However, vibration limits can be expressed in terms of vibration severity measured as the largest orthogonal component of vibration determined in the structure, as shown below.

V_{rms}	Effect
up to 5 mm/s	Damage most unlikely
5–10 mm/s	Damage unlikely
over 10 mm/s	Damage possible – check dynamic stress

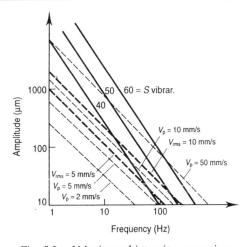

Fig. 5.2.   Velocity and intensity comparison.

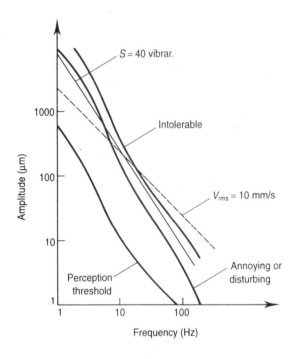

Fig. 5.3.   Human response to vibration.

Although it is essential for the dynamic stresses and strains in a structure to be withstood by the components of the structure, and that failure due to fatigue or malfunctioning must not occur, in many structures such as vehicles and buildings the response of people to the expected vibration must be considered. Human perception of vibration is very good, so that it is often a real challenge in structural design to ensure that the perception threshold level is not exceeded. An indication of the likely human response to vertical vibration is shown in Fig. 5.3 together with the lines for $V_{rms} = 10$ mm/s and $S = 40$ vibrar. The threshold for sensing harmonic vibration, both when standing and lying down, can be predicted fairly accurately by using the Diekmann criteria K values as in the table opposite.

The approximate threshold of human vibration perception corresponds quite closely to the Diekmann criteria for $K = 1$. It can be seen, therefore, that $K = 0.1$ is a very conservative and safe value for predicting the perception threshold, which is the vibration level that should not be exceeded in buildings from a human tolerance viewpoint, these levels being well within the dynamic capabilities of a structure; that is, only in those structures which do not have human operators or occupants may vibration levels be such that structural damage may occur. Naturally this excludes special cases such as earthquake excitation.

The Diekmann K values

Vertical vibration:

below 5 Hz	$K = Af^2$
between 5 Hz and 40 Hz	$K = Af$
above 40 Hz	$K = 200A$

Horizontal vibration:

below 2 Hz	$K = 2Af^2$
between 2 Hz and 25 Hz	$K = 4Af$
above 25 Hz	$K = 100A$

where A = amplitude of vibration in mm, and f = frequency in Hz.

The regions for vibration sensitivity are as follows:
$K = 0.1$, lower limit of perception.
$K = 1$, allowable in industry for any period of time.
$K = 10$, allowable for short duration only.
$K = 100$, upper limit of strain for the average man.

5.4 STRUCTURAL DAMAGE

Structures such as offices, factories, bridges, ships and high-rise buildings are subjected to vibrations generated by a number of sources including machinery, vehicles, trains, aircraft and cross winds. A clear distinction must be made between high-intensity short-duration vibration induced by earthquakes and blasting, and the long-duration usually lower-intensity vibrations such as those induced by traffic and machinery. In particular, buildings are more likely to be damaged by strong dynamic loads such as those generated by earthquakes. Subsequent vibration from other sources can then cause existing cracks to develop and the structural stiffness to vary and eventually a resonance may occur. This condition can cause the vibration to increase beyond structurally safe limits. However, the resistance to fatigue of steel and reinforced concrete structures is such that damage is unlikely to occur if the level of vibration can be tolerated by its occupants.

It is often difficult to establish with certainty the cause of damaging vibrations. For example, cracks in buildings may be due to the vibration from underground trains or aircraft, or merely be building settlement following changes in moisture content of the building fabric or foundations.

For industrial buildings and structures damage may be interpreted as a decrease in either their safety state, their load carrying capacity or their ability to fulfil the desired function. For public buildings and homes damage also refers to the initiation of plaster cracks or the development of existing cracks; this damage is usually superficial and can be easily remedied. Nuclear power, gas and chemical plants are particularly vulnerable to structural damage, however, where no possibility of failure or reduction in structural integrity can be allowed because any leakage may be disastrous.

It should be noted that in all cases of long-duration vibration, damage does not refer to building collapse or complete failure. The limit values of allowable vibration provide quite a large safety margin against yielding or failure in a structural or material sense. Maximum allowable steady-state vibration levels are lower than those for shock-induced short-duration vibrations such as those caused by blasting and earthquakes.

5.5 EFFECTS OF DAMPING ON VIBRATION RESPONSE OF STRUCTURES

It is desirable for all structures to possess sufficient damping so that their response to the expected excitation is acceptable. Increasing the damping in a structure will reduce its response to a given excitation. Thus if the damping in a structure is increased there will be a reduction in vibration and noise, and the dynamic stresses in the structure will be reduced with a resulting benefit to the fatigue life. Naturally the converse is also true.

However it should be noted that increasing the damping in a structure is not always easy, it can be expensive and it may be wasteful of energy during normal operating conditions.

Some structures need to possess sufficient damping so that their response to internally generated excitation is controlled: for example, a crane structure has to have a heavily damped response to sudden loads, and machine tools must have adequate damping so that a heavily damped response to internal excitation occurs, so that the cutting tool produces a good and accurate surface finish with a high cutting speed. Other structures such as chimneys and bridges must possess sufficient damping so that their response to external excitation such as cross winds does not produce dynamic stresses likely to cause failure through fatigue. In motor vehicles, buildings and ships, noise and vibration transmission through an inadequately damped structure may be a major consideration.

Before considering methods for increasing the damping in a structure, it is necessary to be able to measure structural damping accurately.

5.6 THE MEASUREMENT OF STRUCTURAL DAMPING

It must be appreciated that in any structure a number of mechanisms contribute to the total damping. Different mechanisms may be significant at different stress levels, temperatures or frequencies. Thus damping is both frequency and mode dependent, both as to its mechanism and its magnitude. In discussing the effect of various variables on the total damping in a structure it is essential therefore to define all the operating conditions.

Sometimes it is not possible to measure the damping occurring in a structure on its own. For example ships have to be tested in water, which significantly effects the total damping. However, since the ship always operates in water, this total damping is relevant; what is not clear is how changes in the structure will affect the total in-water damping; cargos may also have some effect. On the other hand, a structure such as a machine tool can be tested free of liquids and workpiece; indeed the damping of each structural component can be measured in an attempt to find the most significant source of damping, and hence the most efficient way of increasing the total damping in the structure.

In all cases when damping measurements are being carried out a clear idea of exactly what is being measured is essential. It must be noted that in some tests carried out the damping within the test system itself has, unfortunately, been the major contributor to the total damping.

It has been seen in chapter 2 that the free-decay method is a convenient way for assessing the damping in a structure. The structure is set into free vibration by a shock load such as a small explosive; the fundamental mode dominates the response since all the higher modes are damped out quite quickly. It is not usually possible to excite any mode other than the fundamental using this method. By measuring and recording the decay in the oscillation the logarithmic decrement Λ is found, where

$$\Lambda = \ln \left(\frac{\text{amplitude of motion}}{\text{amplitude of motion one cycle later}} \right).$$

If the damping is viscous, or acts in an equivalent viscous manner Λ will be a constant irrespective of the amplitude. To check this, the natural logarithm of the amplitudes can be plotted against cycles of motion; viscous damping gives a straight line, as shown in Fig. 5.4.

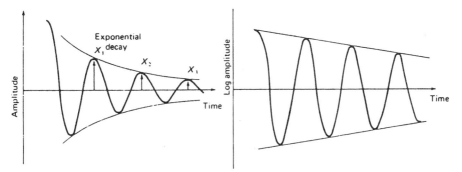

Fig. 5.4.   Vibration decay for viscous damped system.

For viscous damping,

$$\Lambda = \ln \left(\frac{X_1}{X_2} \right) = \ln \left(\frac{X_2}{X_3} \right) = \ln \left(\frac{X_3}{X_4} \right) = \dots = \ln \left(\frac{X_{n-1}}{X_n} \right).$$

Thus

$$n\Lambda = \ln \frac{X_1}{X_2} \cdot \frac{X_2}{X_3} \cdot \frac{X_3}{X_4} \dots \frac{X_{n-2}}{X_{n-1}} \cdot \frac{X_{n-1}}{X_n},$$

that is,

$$\Lambda = \frac{1}{n} \ln \left(\frac{X_1}{X_n} \right),$$

which is a useful expression to use if Λ is small. Note that $\Lambda = 2\pi\zeta/\sqrt{(1 - \zeta^2)}$, and for low damping $\Lambda \simeq 2\pi\zeta$.

There are several ways of expressing the damping in a structure; one of the most common is by the Q factor.

When a structure is forced into resonance by a harmonic exciting force, the ratio of the maximum dynamic displacement at steady-state conditions to the static displacement under a similar force is called the Q factor, that is,

$$Q = \frac{X_{\text{max.dyn.}}}{X_{\text{static}}} = \frac{1}{2\zeta} \text{ (section 2.3.1).}$$

Since a structure can be excited into resonance at any of its modes, a Q factor can be determined for each mode.

Example 38

A single degree of freedom vibrational system of very small viscous damping ($\zeta < 0.1$) is excited by a harmonic force of frequency v and amplitude F. Show that the Q factor of the system is equal to the reciprocal of twice the damping ratio ζ. The Q factor is equal to $(X/X_s)_{\text{max}}$.

It is sometimes difficult to measure ζ in this way because the static deflection X_s of the body under a force F is very small. Another way is to obtain the two frequencies p_1 and p_2 (one either side of the resonance frequency ω) at the half-power points.

Show that $Q = 1/2\zeta = \omega/(p_2 - p_1) = \omega/\Delta\omega$.

(The half-power points are those points on the response curve with an amplitude $1/\sqrt{2}$ times the amplitude at resonance.)

From equation (2.12) above

$$X = \frac{F/k}{\sqrt{\left\{\left[1 - \left(\frac{v}{\omega}\right)^2\right]^2 + \left[2\zeta\frac{v}{\omega}\right]^2\right\}}}.$$

If $v = 0$, $X_s = F/k$, and at resonance $v = \omega$, so

$$X_{\text{max}} = (F/k)/2\zeta = X_s/2\zeta,$$

that is,

$$Q = (X/X_{\text{static}})_{\text{max}} = 1/2\zeta.$$

If X_s cannot be determined, the Q factor can be found by using the half-power point method. This method requires very accurate measurement of the vibration amplitude for excitation frequencies in the region of resonance. Once X_{max} and ω have been located, the so-called half-power points are found when the amplitude is $X_p = X_{\text{max}}/\sqrt{2}$ and the corresponding frequencies either side of ω, p_1 and p_2 determined. Since the energy dissipated per cycle is proportional to X^2, the energy dissipated is reduced by 50% when the amplitude is reduced by a factor of $1/\sqrt{2}$.

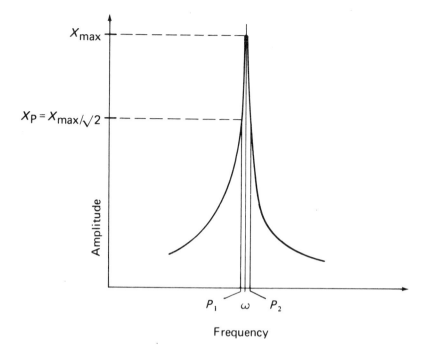

Amplitude–frequency response

Now

$$X = \frac{F/k}{\sqrt{\left\{\left[1 - \left(\dfrac{v}{\omega}\right)^2\right]^2 + \left[2\zeta\dfrac{v}{\omega}\right]^2\right\}}}.$$

Thus

$$X_{max} = \frac{F/k}{2\zeta} \quad \left(\zeta \text{ small, so } X_{max} \text{ occurs at } \frac{v}{\omega} \simeq 1\right)$$

and

$$X_p = \frac{X_{max}}{\sqrt{2}} = \frac{F/k}{\sqrt{2}.2\zeta} = \frac{F/k}{\sqrt{\left\{\left[1 - \left(\dfrac{p}{\omega}\right)^2\right]^2 + \left[2\zeta\dfrac{p}{\omega}\right]^2\right\}}}.$$

Hence

$$\left[1 - \left(\frac{p}{\omega}\right)^2\right]^2 + \left[2\zeta\frac{p}{\omega}\right]^2 = 8\zeta^2$$

and

$$\left(\frac{p}{\omega}\right)^2 = (1 - 2\zeta^2) \pm 2\zeta\sqrt{(1 - \zeta^2)};$$

that is,

$$\frac{p_2{}^2 - p_1{}^2}{\omega^2} = 4\zeta\sqrt{(1 - \zeta^2)} \simeq 4\zeta, \text{ if } \zeta \text{ is small.}$$

Now

$$\frac{p_2{}^2 - p_1{}^2}{\omega^2} = \left(\frac{p_2 - p_1}{\omega}\right)\left(\frac{p_2 + p_1}{\omega}\right) = 2\left(\frac{p_2 - p_1}{\omega}\right),$$

because $(p_1 + p_2)/\omega = 2$; that is, a symmetrical response curve is assumed for small ζ. Thus

$$\frac{p_2 - p_1}{\omega} = 2\zeta = \frac{\Delta\omega}{\omega} = \frac{1}{Q},$$

where $\Delta\omega$ is the frequency bandwidth at the half-power points.

Thus, for light damping, the damping ratio ζ and hence the Q factor associated with any mode of structural vibration can be found from the amplitude–frequency measurements at resonance and the half-power points. Care is needed to ensure that the exciting device does not load the structure and alter the frequency response and the damping, and also that the neighbouring modes do not affect the purity of the mode whose resonance response is being measured.

Some difficulty is often encountered in measuring X_{max} accurately. Fig. 5.5 shows a response in which mode 1 is difficult to measure accurately because of the low damping, that is, the high Q factor. It is difficult to assess the peak amplitude and hence there may be significant errors in the half-power points' location, and a large percentage error in $\Delta\omega$ because it is so small. Measurements for mode 2 would probably give an acceptable value for the Q factor for this mode, but modes 3 and 4 are so close together that they interfere with each other, and the half-power points cannot be accurately found from this data.

In real systems and structures, a very high Q at a low frequency, or a very low Q at a high frequency, seldom occur, but it can be appreciated from the above that very real measuring difficulties can be encountered when trying to measure bandwidths of only a few Hz accurately, even if the amplitude of vibration can be determined. The following table shows the relationship between Q and Δf for different values of frequency.

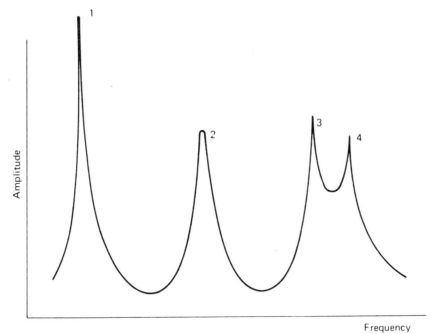

Fig. 5.5.   Amplitude–frequency response, multi-resonance.

Resonance frequency (Hz)	Frequency bandwidth (Hz) for Q factor		
	500	50	5
10	0.02	0.2	2
100	0.2	2	20
1000	2	20	200

An improvement in accuracy in determining Q can often be obtained by measuring both amplitude and phase of the response for a range of exciting frequencies. Consider a single degree of freedom system under forced excitation Fe^{jvt}. The equation of motion is

$$m\ddot{x} + c\dot{x} + kx = Fe^{jvt}.$$

A solution $x = Xe^{jvt}$ can be assumed, so that

$$-mv^2X + jcvX + kX = F.$$

Hence

$$\frac{X}{F} = \frac{1}{(k - mv^2) + jcv}$$

$$= \frac{k - mv^2}{(k - mv^2)^2 + (cv)^2} - j\frac{cv}{(k - mv^2)^2 + (cv)^2};$$

that is, X/F consists of two vectors, $\mathrm{Re}(X/F)$ in phase with the force, and $\mathrm{Im}(X/F)$ in quadrature with the force. The locus of the end point of vector X/F as v varies is shown in Fig. 5.6 for a given value of c. This is obtained by calculating real and imaginary components of X/F for a range of frequencies.

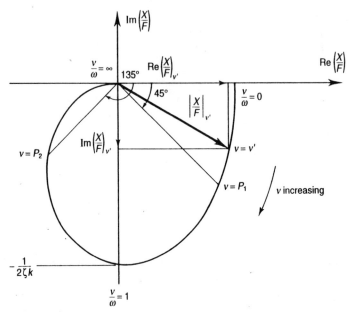

Fig. 5.6. Receptance vector locus for system with viscous damping.

Experimentally this curve can be obtained by plotting the measured amplitude and phase of (X/F) for each exciting frequency.
Since

$$\tan \phi = \frac{k - mv^2}{cv},$$

when $\phi = 45°$ and $135°$,

$$1 = \frac{k - mp_1^2}{cp_1} \quad \text{and} \quad -1 = \frac{k - mp_2^2}{cp_2}.$$

Hence

$$mp_1^2 + cp_1 - k = 0 \text{ and } mp_2^2 - cp_2 - k = 0.$$

Subtracting one equation from the other gives $p_2 - p_1 = c/m$, or

$$\frac{p_2 - p_1}{\omega} = \frac{\Delta\omega}{\omega} = 2\zeta = \frac{1}{Q};$$

that is, X/F at resonance lies along the imaginary axis, and the half-power points occur when $\phi = 45°$ and $135°$. If experimental results are plotted on these axes a smooth curve can be drawn through them so that the half-power points can be accurately located.

The method is also effective when the damping is hysteretic, because in this case

$$\frac{X}{F} = \frac{1}{(k - m v^2) + j\eta k},$$

so that

$$\mathrm{Re}\left(\frac{X}{F}\right) = \frac{k - m v^2}{(k - m v^2)^2 + (\eta k)^2} \quad \text{and} \quad \mathrm{Im}\left(\frac{X}{F}\right) = \frac{-\eta k}{(k - m v^2)^2 + (\eta k)^2}.$$

Thus

$$\left[\mathrm{Re}\left(\frac{X}{F}\right)\right]^2 + \left[\mathrm{Im}\left(\frac{X}{F}\right)\right]^2 = \frac{1}{(k - m v^2)^2 + (\eta k)^2}$$

or

$$\left[\mathrm{Re}\left(\frac{X}{F}\right)\right]^2 + \left[\mathrm{Im}\left(\frac{X}{F}\right) - \frac{1}{2\eta k}\right]^2 = \left(\frac{1}{2\eta k}\right)^2;$$

that is, the locus of (X/F) as v increases from zero is part of a circle, centre $(0, -1/2\eta k)$ and radius $1/2\eta k$, as shown in Fig. 5.7.

In this case, therefore, it is particularly easy to draw an accurate locus from a few experimental results, and p_1 and p_2 are located on the horizontal diameter of the circle.

This technique is known variously as a frequency locus plot, Kennedy–Pancu diagram or Nyquist diagram.

It must be realized that the assessment of damping can only be approximate. It is difficult to obtain accurate, reliable, experimental data, particularly in the region of resonance; the analysis will depend upon whether viscous or hysteretic damping is assumed, and some non-linearity may occur in a real system. These effects may cause the frequency-locus plot to rotate and translate in the $\mathrm{Re}(X/F)$, $\mathrm{Im}(X/F)$ plane. In these cases the resonance frequency can be found from that part of the plot where the greatest rate of change of phase with frequency occurs (Figs 2.24 and 2.25).

5.7 SOURCES OF DAMPING

The damping which occurs in structures can be considered to be either inherent damping, that is, damping which occurs naturally within the structure or its environment, or added damping which is that resulting from specially constructed dampers added to the structure.

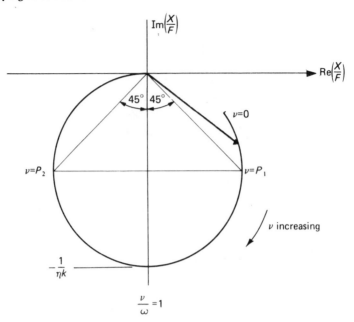

Fig. 5.7.   Receptance vector locus for system with hysteretic damping.

5.7.1 Inherent damping

5.7.1.1 Hysteretic or material damping

All materials dissipate some energy during cyclic deformation. The amount may be very small, however, and is linked to mechanisms associated with internal reconstruction such as molecular dislocations and stress changes at grain boundaries. Such damping effects are non-linear and variable within a material so that the analysis of such damping mechanisms is difficult. However, experimental measurements of the behaviour of samples of specific materials can be made to determine the energy dissipated for various strain levels.

For most conventional structural materials the energy dissipated is very small. Because of this the actual damping mechanisms within a given material are usually of limited interest, particularly in view of the uncertainty of describing the actual mechanisms and the difficulty with carrying out a reasonable theoretical analysis. However, some particular materials, which are known as high damping alloys, have been developed which have had a certain damping mechanism enhanced (see section 5.7.2.1 below).

In order to determine the energy dissipated within a material, hysteresis load extension loops are usually plotted. The load extension hysteresis loops for linear materials and structures are elliptical under sinusoidal loading, and increase in area according to the square of the extension. Although the loss factor η of a material depends upon its composition, temperature, stress and the type of loading mechanism used, an approximate value for η can be obtained. It should be noted that the deviation of these loops from a single line is usually very small so that damping arising from the constituent material of a structure is usually very small also and may be insignificant compared with the other damping mechanisms in a structure.

A range of values of η for some common engineering materials is given in the table. For more detailed information on material damping mechanisms and loss factors, see *Damping of Materials and Members in Structural Mechanisms* by B. J. Lazan (Pergamon).

Material	Loss factor
Aluminium – pure	0.00002–0.002
Aluminium alloy – dural	0.0004–0.001
Steel	0.001–0.008
Lead	0.008–0.014
Cast iron	0.003–0.03
Manganese copper alloy	0.05–0.1
Rubber – natural	0.1–0.3
Rubber – hard	1.0
Glass	0.0006–0.002
Concrete	0.01–0.06

The high damping metals and alloys referred to above are often unsuitable for engineering structures because of their low strength, ductility and hardness, and their high cost. Manganese copper is an exception in that it has high ultimate tensile strength, hardness and ductility. However, these special alloys are difficult to produce, and their damping is only large at high strains which means that structures have to endure high vibration levels which may lead to other problems such as fatigue or excessive noise. It will be realized that a steel or aluminium structure with material damping alone for which $\eta = 0.001$, will have a Q factor of the order of 1000. This would be unacceptable in practice and fortunately rarely arises because of the significant additional damping that occurs in the structural joints.

5.7.1.2 Damping in structural joints

The Q factor of a bolted steel structure is usually between 20 and 60, and for a welded steel structure a Q factor between 30 and 100 is common. Reinforced concrete can have Q factors in the range of 15 to 25. Since the damping in the structural material is very small, most of the damping which occurs in real structures arises in the structural joints. However, even though over 90% of the inherent damping in most structures arises in the structural joints, little effort is made to optimize or even control this source of damping. This is because the energy dissipation mechanism in a joint is a complex process which is largely influenced by the interface pressure. At low joint clamping pressures sliding on a macro scale takes place and Coulomb's Law of Friction is assumed to hold. If the joint clamping pressure increases, mutual embedding of the surfaces starts to occur. Sliding on a macro-scale is reduced and micro-slip is initiated which involves very small displacements of an asperity relative to its opposite surface. A further increase in the joint clamping pressure will cause greater penetration of the asperities. The pressure on the contact areas will be the yield pressure of the softer material. Relative motion causes further plastic deformation of the asperities.

In most joints all three mechanisms operate, their relative significance depending upon the joint conditions. In joints with high normal interface pressures and relatively rough surfaces, the plastic deformation is significant. Many joints have to carry pressures of this magnitude to satisfy criteria such as high static stiffness. A low normal interface pressure would tend to increase the significance of the slip mechanisms, as would an improvement in the quality of the surfaces in contact. With the macro-slip mechanism, the energy dissipation is proportional to the product of the interface shear force and the relative tangential motion. Under high pressure, the slip is small, and under low pressure the shear force is small: between these two extremes, the product becomes a maximum.

However, when two surfaces nominally at rest with respect to each other are subjected to slight vibrational slip, fretting corrosion can be instigated. This is a particularly serious form of wear inseparable from energy dissipation by interfacial slip, and hence frictional damping.

The fear of fretting corrosion occurring in a structural joint is one of the main reasons why joints are tightly fastened. However, joint surface preparations such as cyanide hardening and electro-discharge machining are available which reduce fretting corrosion from frictional damping in joints considerably, whilst allowing high joint damping. Plastic layers and greases have been used to separate the interfaces in joints and prevent fretting, but they have been squeezed out and have not been durable. Careful joint design and location is necessary if joint damping is to be increased in a structure without fretting corrosion becoming a problem; full details of fretting are given in *Fretting Fatigue* by R. B. Waterhouse (Applied Science Publishers, 1974).

The theoretical assessment of the damping that may occur in joints is difficult to make because of the variations in μ that occur in practice. However, it is generally accepted that the friction force generated between the joint interfaces is usually:

(i)   dependent on the materials in contact and their surface preparation;
(ii)   proportional to the normal force across the interface;
(iii) substantially independent of the sliding speed and apparent area of contact;
(iv)   greater just prior to the occurrence of relative motion than during uniform relative motion.

The equations of motion of a structure with friction damping are thus non-linear: most attempts at analysis linearize the equations in some way. A very useful method is to calculate an equivalent viscous damping coefficient such that the energy dissipated by the friction and viscous dampers is the same. This has been shown to give an acceptable qualitative analysis for macro-slip. Some improvement on this method can be obtained by replacing μ by a term that allows for changes in the coefficient with slip amplitude. Some success has also been obtained by simply replacing the friction force with an equivalent harmonic force which is, essentially, the first term of the Fourier series representing the periodic friction force.

Some effects of controlling the joint clamping forces in a structure can be seen by considering an elastically supported beam fitted with friction joints at each end, as shown in Fig. 5.8.

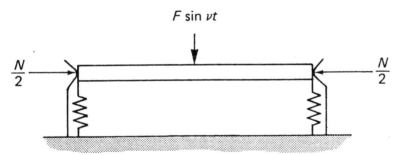

Fig. 5.8.   Elastically supported beam with Coulomb damping.

The beam is excited by the harmonic force F sin vt applied at mid-span. When the friction joints are very slack, $N \simeq 0$ and the beam responds as an elastic beam on spring supports. When the joints are very tight, $N \simeq \infty$ and the beam responds to excitation as if built-in at each end. For $\infty > N > 0$, a damped response occurs such as that shown in Fig. 5.9.

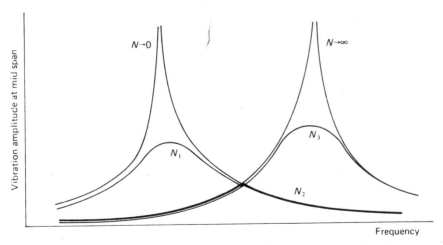

Fig. 5.9.   Amplitude–frequency response for the beam shown in Fig. 5.8.

If N is increased from zero to N_1, a damped elastic response is achieved; significant damping occurs only when the beam vibration is sufficient to cause relative slip in the joints. As N approaches N_2, the beam responds as if built-in, until a vibration amplitude is reached when the joints slip, and the response is the same as that for the damped-elastic beam. When N increases to N_3, the built-in beam response is maintained until a higher amplitude is reached before slip takes place. The minimum response is achieved when $N = N_2$. This is obviously a powerful technique for controlling the dynamic response of structures, since both the maximum response, and the frequency at which this occurs, can be optimized. Friction damping can also be applied to joints that slip in rotation as well as, or instead of, translation.

The damping in plate type structures and elements can be increased by fabricating the plate out of several laminates bolted or riveted together, so that as the plate vibrates interfacial slip occurs between the laminates thus giving rise to frictional damping. A Q factor as low as 20 has been obtained for a freely supported laminated circular plate, produced by clamping two identical plates together to form a plate subjected to interfacial friction forces. For a solid plate, in which only material damping occurred, the Q factor was 1300. Theoretically a laminated plate can be modelled by a single plate subjected to in-plane shear forces. When tightly fastened along two edges a Q factor of 345 was obtained for a square steel plate; adjusting the edge clamping to the optimum allowed the Q factor to fall to 15, for the first mode of vibration. Replacing the plate by two similar plates, each half the thickness of the original enabled a Q factor of 75 to be achieved, even when the edges were tightly clamped. This improved to a Q of 25 when optimum edge clamping was applied. However, some loss in stiffness must be expected, leading to a reduction in the resonance frequencies.

This technique has been applied with some success to plate type structural elements such as engine oil sumps, for reducing the noise and vibration generated.

It is often unnecessary to add a special damping device to a structure to increase the frictional damping, optimization of an existing joint or joints being all that is required. Thus it can be cheap and easy to increase the inherent damping in a structure by optimizing the damping in joints, although careful design is sometimes necessary to ensure that adequate stiffness is maintained. It must be recognized that for joint damping to be large, slip must occur, and that fretting corrosion and joint damping are inseparable. Furthermore, some of the stiffness of a tightly clamped structure must be sacrificed if this source of damping is to be increased, although this loss in stiffness need not be large if the joints are carefully selected. This damping mechanism is most effective at low frequencies and the first few modes of vibration, since only under these conditions are the vibration amplitudes generally large enough to allow significant slip, and therefore damping, from this mechanism.

Notwithstanding the difficulties of analysis and the application of the damping in structural joints, some form of this damping occurs in all structures. It is rarely used efficiently, optimized or even controlled however, but it does have useful advantages so that it deserves wider application. There is a wide range of dynamic systems and structures that would benefit from increased joint damping such as beam systems, frameworks, gas turbines and aerospace structures.

5.7.1.3 Acoustic radiation damping

The vibrational motions of a structure will always couple with the surrounding fluid medium, such as air or water, so that its response is affected. Generally this effect is very small so that this source of damping is not usually large enough to be useful. There are exceptions, however, such as aircraft panels constructed from thin lightweight stiffened structures, but for heavy machines and structures air is much too thin to exert any significant pressure on the vibrating surfaces so that the damping from this source is negligible. It should be appreciated that acoustic damping cannot occur in spacecraft or other structures in a similar environment.

The damping effect of the surrounding fluid medium depends upon a number of parameters such as the medium density and the mass and stiffness of the structure.

Accordingly, acoustic radiation damping is much higher in water or in oil than it is in air and this type of damping is far more effective for high frequencies than low. It should be noted that acoustic pressures from some parts of a vibrating structure may cancel out those from other parts, for example when modes of vibration are in antiphase, so that acoustic damping can be disappointingly very small.

The analysis of acoustic damping often leads to very complicated formulae which are difficult to evaluate except in specific cases. However, some theoretical estimate can usually be made if it is considered that this form of damping, that is the radiation of vibrational energy in the form of sound waves within the surrounding medium, could be significant within a given application.

5.7.1.4 Air pumping

Consider a part of a fabricated structure, such as a panel which is vibrating. If a cover or adjacent member of different relative stiffness and vibration characteristics is attached, as shown in Fig. 5.10, then during vibration the volume of the enclosed space changes. If some sort of opening is provided, either by chance or by design, air will be pumped through the leakage holes. The air flow may be laminar or turbulent depending upon the amplitude of vibration, the enclosed volume, the size of the leakage hole, mode of vibration and so on. For some panel modes of vibration, this flow may be very small; this is particularly true for high-frequency modes with nodal lines within the cavity so that some parts of the panel surface motion are out of phase with others. If the flow is small, it follows that the damping will be small. Damping from air pumping at high frequencies tends to be very low, therefore, and in addition it is found to be inversely proportional to the frequency squared. It should also be noted that the flow paths are difficult to determine so that the damping that occurs from air pumping is correspondingly difficult to evaluate.

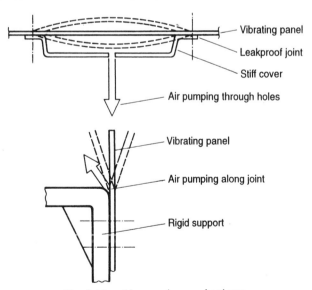

Fig. 5.10.   Air pumping mechanisms.

However, it is necessary to be aware of its existence and significance. In a particular structure this form of damping can be evaluated by testing in air and also in a vacuum when the pumping action effects will be zero.

5.7.1.5 Aerodynamic damping

Energy can be dissipated by the air in which a structure vibrates. This can be important for low-density structures with large motions. Most damping forces are of a retarding nature which act against the motion occurring, but situations can arise when the motion itself generates a force that encourages motion. When this happens in a structure due to relative motion of the wind, negative aerodynamic damping or aerodynamic instability occurs. Of course aerodynamic damping can be positive but motion instability is often associated with aerodynamic effects.

There are several methods of aerodynamic excitation, which may be considered to be negative damping, which induce structural vibration, such as buffeting by wind eddies or wake turbulence from an upstream body. For many structures there is insufficient wind energy to excite significant vibration but in steady cross winds vortex generation can cause galloping, aeolian vibration and flutter. Galloping is the large-amplitude low-frequency oscillation of long cylindrical structures exposed to a transverse wind; it is frequently observed on overhead power lines. Aeolian vibration, which also occurs on overhead power lines, is a higher-frequency oscillation which arises from vortex shedding in steady cross winds. Flutter is a motion that relies on the aerodynamic and inertial coupling between two modes of vibration. Structures commonly affected are suspension bridges and tall non-circular towers and stacks where substantial bending and torsion occur. Aerodynamic excitation by vortex shedding is probably the most common of all wind-induced vibrations as discussed in section 2.3.7.

Wind forces on buildings and structures are always unsteady and may be due to variations in the wind gusts, vortex shedding or the interaction between the inertial, elastic and aerodynamic forces. The most dangerous unsteady forces are those that are cyclic since the frequency of the fluctuating part may coincide with a natural frequency.

In the design of tall slender structures such as chimneys, stacks and towers, it is essential that the natural frequencies of the bending and torsion modes are well separated from the vortex shedding frequencies. This often means modifying the structure to alter the vortex shedding; this is usually done by adding helical strakes to the top quarter of the structure. The damping of a steel stack can also be increased by applying coatings or additional dampers. Tall chimneys with several flues can be perforated to relieve pressure differentials. Aeolian vibration of overhead powerlines is usually controlled by fitting a damped vibration absorber (section 5.7.2.5); failure to damp these vibrations adequately leads to fatigue failure.

5.7.1.6 Other damping sources

In general the major sources of damping in a structure are within the joints and the structural material. Occasionally, however, structures are required to work in environments that contribute significantly to the total damping. For example, ship hulls benefit from the considerable hydrodynamic damping of the water: this is true for all water-immersed structures; and aerodynamic damping, though itself small, may be important in lightly damped structures.

5.7.2 Added damping

When the inherent damping in a structure is insufficient, it can be increased either by adding vibration dampers to the structure or by manufacturing the structure, or a part of it, out of a layered material with very high damping properties.

5.7.2.1. High damping alloys

From the discussion in section 5.7.1.1. on material damping it can be deduced that unless the effect of the damping mechanisms within a given material can be deliberately increased, the material damping effects on the response of a structure or dynamic system will be very small indeed. To this end, particular alloy materials have been developed which are such that their structure allows increased damping within the material. Unfortunately this gain in damping is often at the expense of other desirable material properties such as stiffness, strength, machinability and cost, so that these materials themselves are not usually suitable for structural purposes. Sometimes, however, situations arise when the use of such materials can be beneficial, as in aerospace structures. Because of the highly non-linear behaviour of these materials their damping is best evaluated experimentally in terms of modal damping and natural frequencies.

5.7.2.2. Composite materials

A composite material is usually considered to be one which is a combination of two or more constituent materials on a macroscopically homogeneous level. Examples of such composites are an aluminium matrix embedded with boron fibres and an epoxy matrix embedded with carbon fibres. The fibres may be long or short, directionally aligned or randomly orientated, or some sort of mixture, depending on the intended use of the material. Unconventional manufacturing and construction techniques are usually necessary.

The objective is to increase the stiffness and at the same time reduce the weight of a structure. This naturally has some effect on the dynamic properties both as regards natural frequencies and damping. Composite materials are usually expensive so their application is often linked to critical areas of a structure such as parts of an aircraft fuselage or wing, space vehicles and racing car shells. There are disadvantages, however, such as their low resistance to erosion, high cost and repair difficulties.

Although not developed for their damping properties, composite materials can possess high damping. This occurs when stiff fibres are embedded in a highly damped matrix material. The fibres give the necessary strength and stiffness properties and the matrix provides the damping. Particular care should be taken when testing these materials to distinguish between the effects of damping and non-linearities (section 5.9).

5.7.2.3. Viscoelastic materials

Viscoelastic damping occurs in many polymers and this internal damping mechanism is widely used in structures and machines for controlling vibration (Fig. 5.11). The damping arises from the polymer network after it has been deformed. Both frequency and temperature effects have a large bearing on the molecular motion and hence on the damping characteristics.

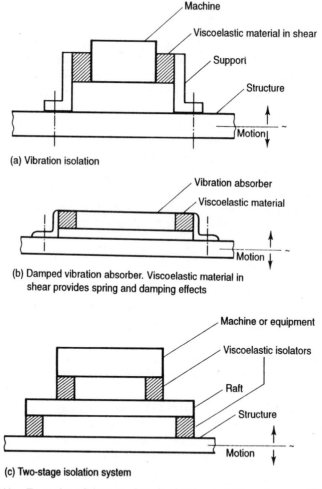

(a) Vibration isolation

(b) Damped vibration absorber. Viscoelastic material in
shear provides spring and damping effects

(c) Two-stage isolation system

Fig. 5.11.   Examples of the use of viscoelastic material to reduce machine and
structure vibration.

With careful control, polymer materials can be manufactured with a wide range of properties such as high damping, strength and good creep resistance over a useful range of temperatures and frequencies. They often feature in antivibration mountings and as the constrained layer material in highly damped composite beams and plates.

In addition it should be noted that a common method of applying damping to a plate-type structural element or panel vibrating in a bending mode is to spray the surface with a layer of viscoelastic material possessing high internal losses. The most well-known materials that are specially made for this purpose are the mastic deadeners made using an asphalt base. The ratio of the thickness of the damping layer to the thickness of the structure is very important and is usually between one and three. One thick single-sided layer of material is more effective than two thin double-sided ones.

When designing systems using viscoelastic damping materials it must be appreciated that the static stiffness is usually much less than the dynamic stiffness. The experimental determination of the stiffness and damping properties must take this into account, together with any static preload.

5.7.2.4 Constrained layer damping

The polymer materials that exist with very high damping properties lack sufficient rigidity and creep resistance to enable a structure to be fabricated from them, so that if advantage is to be taken of their high damping a composite construction of a rigid material, such as a metal, with damping layers bonded to it has to be used, usually as a beam or plate. High damping material can be applied to a structure by fabricating it, at least in part, from elements in which layers of high damping viscoelastic material are bonded between layers of metal. When the composite material vibrates the constrained damping layers are subjected to shear effects, which cause vibrational energy to be converted into heat and hence dissipated. Other applications of high damping polymers are to edge damping, where the polymer forms the connection between a panel or beam and its support, and unconstrained layers, where the damping material is simply bonded to the surface of the structural element. Whilst these applications do increase the total damping, they are not as effective as using constrained layers.

Before considering the damping effects that can be achieved by the constrained layer technique, it must be emphasized that the properties of viscoelastic materials are both temperature- and frequency-sensitive. Fig. 5.12 shows how shear modulus and loss factor can vary.

Another disadvantage with composite materials is that they are difficult to bend or form without reducing their damping capabilities, because of the distortion that occurs in the damping layer.

Two, three, four and five or more layers of viscoelastic material and metal can be used in a composite; each layer can have particular properties, thickness and location relative to the neutral axis so that the composite as a whole has the most desirable structural and dynamic performance. Because of this wide variation in composite material geometry, only a three-layer symmetrical construction will be considered, other geometries being an extension of the three-layer composite.

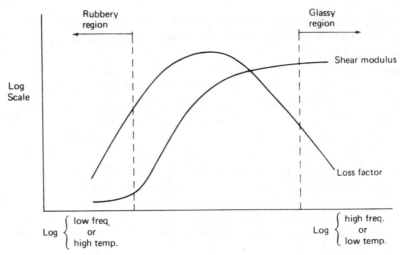

Fig. 5.12. Viscoelastic material properties.

Consider the composite beam of length l shown in Fig. 5.13. A dimensionless shear parameter can be defined, equal to

$$\frac{2l^2 G}{h_1 h_2 E}.$$

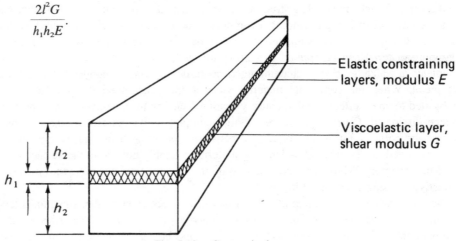

Fig. 5.13. Composite beam.

It is assumed that the elastic constraining layers have a zero loss factor, and that the viscoelastic damping layer has zero stiffness. The beam loss factor for a cantilever vibrating in its first mode is shown as a function of the shear parameter in Fig. 5.14, for various values of the loss factor η for the viscoelastic material.

It can be seen from Fig. 5.14 that a high beam loss factor is only obtainable at a particular value of the shear parameter, and that as the loss factor of the viscoelastic layer increases the curves become sharper. The dependence of the beam loss factor on the shear parameter is consequently of great practical significance. However, very high beam loss factors can be obtained resulting in a Q value of two or even less.

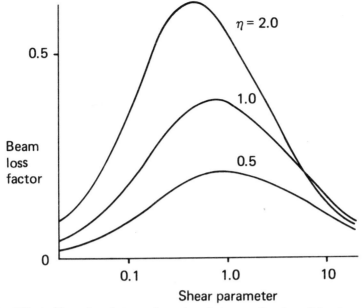

Fig. 5.14.   Effect of layer loss factor on beam loss factor as a function of the shear parameter.

The difference between the optimum loss factors for the first three modes of a cantilever has been shown to be less than 10%. Most viscoelastic materials have a shear modulus which increases with frequency, so that the damping can be kept near to the optimum over a large frequency range. It must be emphasized that it is not possible to secure high structural damping and high stiffness by this method of damping.

Damping in structures, and constrained layer damping in particular, has been discussed in *Structural Damping* by J E Ruzicka (Pergamon Press, 1962), and in *Damping Applications for Vibration Control* edited by P J Torvik (ASME Publication AMD, vol. 38, 1974).

More recently, the damping that can be achieved in structures has been comprehensively studied and researched, particularly with regard to industrial, military and aerospace applications, and improving analytical techniques. The results of some of this work have been published in conference proceedings and relevant learned journals such as *The Journal of Sound and Vibration, The Proceedings of the ASME* and *The Shock and Vibration Digest*.

5.7.2.5 Vibration dampers and absorbers

A wide range of damping devices is commercially available; these may rely on viscous, dry friction or hysteretic effects. In most cases some degree of adjustment is provided, although the effect of the damper can usually be fairly well predicted by using the above theory. The viscous type damper is usually a cylinder with a closely fitting piston and filled with a fluid. Suitable valves and porting give the required resistance to motion of the piston in the cylinder. Dry friction dampers rely on the friction force generated between two or more surfaces pressed together under a controlled force. Hysteretic type dampers are usually made from an elastic material with high internal damping, such as natural rubber. Occasionally dampers relying on other effects such as eddy currents are used.

However, these added dampers only act to reduce the vibration of a structure. If a particularly troublesome resonance exists it may be preferable to add a *vibration absorber*. This is simply a spring–body system which is added to the structure; the parameters of the absorber are chosen so that the amplitude of the vibration of the structure is greatly reduced, or even eliminated, at a frequency that is usually chosen to be at the original troublesome resonance.

The undamped dynamic vibration absorber

If a single degree of freedom system or mode of a multi-degree of freedom system is excited into resonance, large amplitudes of vibration result with accompanying high dynamic stresses and noise and fatigue problems. In most mechanical systems this is not acceptable.

If neither the excitation frequency nor the natural frequency can conveniently be altered, this resonance condition can often be successfully controlled by adding a further single degree of freedom system. Consider the model of the system shown in Fig. 5.15 where K and M are the effective stiffness and mass of the primary system when vibrating in the troublesome mode.

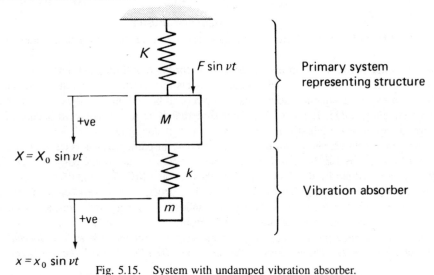

Fig. 5.15.   System with undamped vibration absorber.

The absorber is represented by the system with parameters k and m. From section 3.1.3 it can be seen that the equations of motion are

$$M\ddot{X} = -KX - k(X - x) + F \sin \nu t \quad \text{for the primary system}$$

and

$$m\ddot{x} = k(X - x) \text{ for the vibration absorber.}$$

Substituting

$$X = X_0 \sin \nu t \text{ and } x = x_0 \sin \nu t$$

gives

$$X_0(K + k - Mv^2) + x_0(-k) = F$$

and

$$X_0(-k) + x_0(k - mv^2) = 0.$$

Thus

$$X_0 = \frac{F(k - mv^2)}{\Delta},$$

and

$$x_0 = \frac{Fk}{\Delta},$$

where $\Delta = (k - mv^2)(K + k - Mv^2) - k^2$, and $\Delta = 0$ is the frequency equation.

It can be seen that not only does the system now possess two natural frequencies, Ω_1 and Ω_2 instead of one, but by arranging for $k - mv^2 = 0$, X_0 can be made zero.

Thus if $\sqrt{(k/m)} = \sqrt{(K/M)}$, the response of the primary system at its original resonance frequency can be made zero. This is the usual tuning arrangement for an undamped absorber because the resonance problem in the primary system is only severe when $v \simeq \sqrt{(K/M)}$ rad/s. This is shown in Fig. 5.16.

When $X_0 = 0$, $x_0 = -F/k$, so that the force in the absorber spring, kx_0 is $-F$; thus the absorber applies a force to the primary system which is equal and opposite to the exciting force. Hence the body in the primary system has a net zero exciting force acting on it and therefore zero vibration amplitude.

If an absorber is correctly tuned, $\omega^2 = K/M = k/m$, and if the mass ratio $\mu = m/M$, the frequency equation $\Delta = 0$ is

$$\left(\frac{v}{\omega}\right)^4 - (2 + \mu)\left(\frac{v}{\omega}\right)^2 + 1 = 0.$$

This is a quadratic equation in $(v/\omega)^2$. Hence

$$\left(\frac{v}{\omega}\right)^2 = \left(1 + \frac{\mu}{2}\right) \pm \sqrt{\left(\mu + \frac{\mu^2}{4}\right)},$$

and the natural frequencies Ω_1 and Ω_2 are found to be

$$\frac{\Omega_{1,2}}{\omega} = \left[\left(1 + \frac{\mu}{2}\right) \pm \sqrt{\left(\mu + \frac{\mu^2}{4}\right)}\right]^{1/2}.$$

For a small μ, Ω_1 and Ω_2 are very close to each other, and near to ω; increasing μ gives better separation between Ω_1 and Ω_2 as shown in Fig. 5.17.

This effect is of great importance in those systems where the excitation frequency may vary; if μ is small, resonances at Ω_1 or Ω_2 may be excited. It should be noted that since

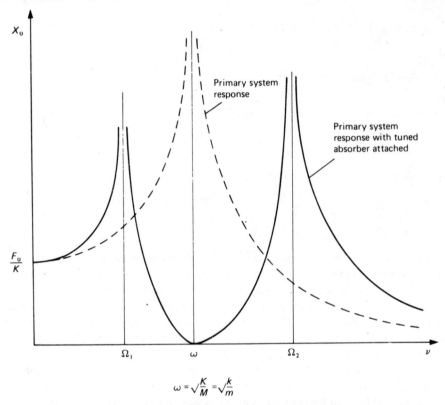

$$\omega = \sqrt{\frac{K}{M}} = \sqrt{\frac{k}{m}}$$

Fig. 5.16. Amplitude–frequency response for system with and without tuned absorber.

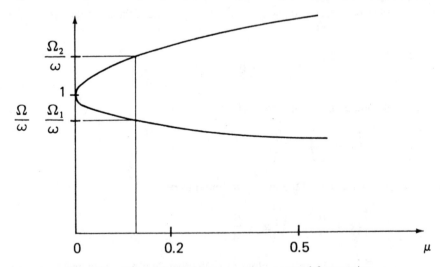

Fig. 5.17. Effect of absorber mass ratio on natural frequencies.

$$\left(\frac{\Omega_1}{\omega}\right)^2 = \left(1 + \frac{\mu}{2}\right) - \sqrt{\left(\mu + \frac{\mu^2}{4}\right)}$$

and

$$\left(\frac{\Omega_2}{\omega}\right)^2 = \left(1 + \frac{\mu}{2}\right) + \sqrt{\left(\mu + \frac{\mu^2}{4}\right)}, \text{ then multiplication gives}$$

$$\frac{\Omega_1{}^2\Omega_2{}^2}{\omega^4} = \left(1 + \frac{\mu}{2}\right)^2 - \left(\mu + \frac{\mu^2}{4}\right) = 1,$$

that is,

$$\Omega_1.\Omega_2 = \omega^2.$$

Also

$$\left(\frac{\Omega_1}{\omega}\right)^2 + \left(\frac{\Omega_2}{\omega}\right)^2 = 2 + \mu.$$

These relationships are very useful when designing absorbers. If the proximity of Ω_1 and Ω_2 to ω is likely to be a hazard, damping can be added in parallel with the absorber spring, to limit the response at these frequencies. Unfortunately, if damping is added, the response at frequency ω will no longer be zero. A design criterion that has to be carefully considered is the possible fatigue and failure of the absorber spring: this could have severe consequences. In view of this, some damped absorbed systems dispense with the absorber spring and sacrifice some of the absorber effectiveness. This is particularly important in torsional systems, where the device is known as a *Lanchester damper*.

Example 39

The figure represents a pump of mass m_1 which rests on springs of stiffness k_1, so that only vertical motion can occur. Given that the damping is negligible and the mass m_2 is ignored, derive an expression for the frequency of the harmonic disturbing force at which the pump will execute vertical oscillations of very large – theoretically infinite – amplitude.

Given that an undamped dynamic absorber of mass m_2 is then connected to the pump by a spring of stiffness k_2, as shown, prove that the amplitude of the oscillations of the pump is reduced to zero when

$$\frac{k_2}{m_2} = v^2,$$

where v is the natural frequency of the free vibrations of the pump in the absence of the dynamic vibration absorber.

The pump has a mass of 130 kg and rotates at a constant speed of 2400 rev/min but due to a rotating unbalance very large amplitudes of pump vibration on the spring supports result. An undamped vibration absorber is to be fitted so that the nearest natural frequency of the system is at least 20% removed from the running speed of 2400 rev/min. Find the smallest absorber mass necessary and the corresponding spring stiffness.

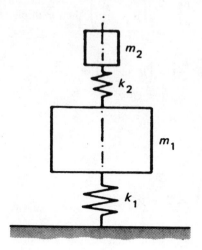

The pump can be modelled as below:

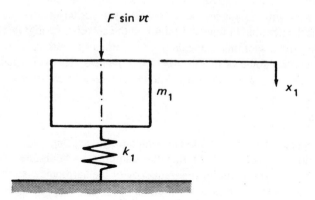

The equation of motion is

$$m_1\ddot{x}_1 + k_1x_1 = F \sin vt,$$

so that if

$$x_1 = X_1 \sin vt, \quad X_1 = \frac{F}{k_1 - m_1v^2}.$$

When

$$v = \sqrt{\left(\frac{k_1}{m_1}\right)}, \quad X_1 = \infty;$$

that is, resonance occurs when $v = \omega = \sqrt{(k_1/m_1)}$. With a vibration absorber added, the system is

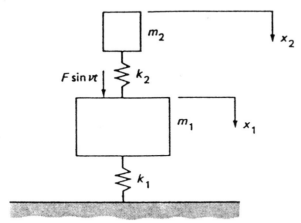

The FBDs are therefore, if $x_2 > x_1$ is assumed,

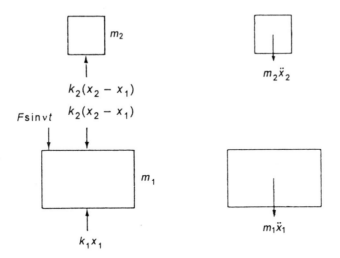

The equations of motion are thus

$$m_2\ddot{x}_2 = -k_2(x_2 - x_1)$$

or

$$m_2\ddot{x}_2 + k_2x_2 - k_2x_1 = 0,$$

and
$$m_1\ddot{x}_1 = k_2(x_2 - x_1) - k_1x_1 + F \sin \nu t$$

or
$$m_1\ddot{x}_1 + (k_1 + k_2)x_1 - k_2x_2 = F \sin \nu t.$$

Assuming $x_1 = X_1 \sin \nu t$ and $x_2 = X_2 \sin \nu t$, these equations give
$$X_1[k_1 + k_2 - m_1\nu^2] + X_2[- k_2] = F$$

and
$$X_1[- k_2] + X_2[k_2 - m_2\nu^2] = 0;$$

that is
$$X_1 = \frac{F(k_2 - m_2\nu^2)}{[(k_1 + k_2) - m_1\nu^2][k_2 - m_2\nu^2] - k_2^2}.$$

Thus

$$\text{if } \nu^2 = \frac{k_2}{m_2}, \quad X_1 = 0.$$

Now the frequency equation is
$$[(k_1 + k_2) - m_1\nu^2][k_2 - m_2\nu^2] - k_2^2 = 0.$$

If we put

$$\mu = \frac{m_2}{m_1} = \frac{k_2}{k_1} \quad \text{and} \quad \Omega = \sqrt{\left(\frac{k_2}{m_2}\right)} = \sqrt{\left(\frac{k_1}{m_1}\right)},$$

this becomes
$$\nu^4 - \nu^2(\mu\Omega^2 + 2\Omega^2) + \Omega^4 = 0$$

or

$$\left(\frac{\nu}{\Omega}\right)^4 - \left(\frac{\nu}{\Omega}\right)^2 (2 + \mu) + 1 = 0,$$

so that

$$\left(\frac{\nu}{\Omega}\right)^2 = \frac{2 + \mu}{2} \pm \sqrt{\left(\frac{\mu^2 + 4\mu}{4}\right)}.$$

The limiting condition for the smallest absorber mass is $(\nu_1/\Omega) = 0.8$ because then $(\nu_2/\Omega) = 1.25$, which is acceptable. Thus

$$\left(0.64 - \frac{\mu + 2}{2}\right)^2 = \frac{\mu^2 + 4\mu}{4}$$

and

$$\mu = 0.2.$$

Hence

$$m_2 = 0.2 \times 130 = 26 \text{ kg},$$

and

$$k_2 = (80\pi)^2 m_2 = 1642 \text{ kN/m}.$$

Example 40

A system has a violent resonance at 79 Hz. As a trial remedy a vibration absorber is attached which results in a resonance frequency of 65 Hz. How many such absorbers are required if no resonance is to occur between 60 and 120 Hz?

Since

$$\left(\frac{\Omega_1}{\omega}\right)^2 + \left(\frac{\Omega_2}{\omega}\right)^2 = 2 + \mu$$

and

$$\Omega_1 \Omega_2 = \omega^2,$$

in the case of one absorber, with $\omega = 79$ Hz and $\Omega_1 = 65$ Hz,

$$\Omega_2 = \frac{79^2}{65} = 96 \text{ Hz}.$$

Also

$$\left(\frac{65}{79}\right)^2 + \left(\frac{96}{79}\right)^2 = 2 + \mu,$$

so $\mu = 0.154$.

In the case of n absorbers, if

$$\Omega_1 = 60 \text{ Hz}, \quad \Omega_2 = \frac{79^2}{60} = 104 \text{ Hz (too low)}.$$

So require $\Omega_2 = 120$ Hz and then $\Omega_1 = (79^2/120) = 52$ Hz. Hence

$$\left(\frac{52}{79}\right)^2 + \left(\frac{120}{79}\right)^2 = 2 + \mu'.$$

Thus

$$\mu' = 0.74 = n\mu \quad \text{and} \quad n = \frac{0.74}{0.154} = 4.82.$$

Thus five absorbers are required.

Example 41

A machine tool of mass 3000 kg has a large resonance vibration in the vertical direction at 120 Hz. To control this resonance, an undamped vibration absorber of mass 600 kg is fitted tuned to 120 Hz. Find the frequency range in which the amplitude of the machine vibration is less with the absorber fitted than without.

If (X_o) with absorber $= (X_o)$ without absorber,

$$\frac{F(k - mv^2)}{(K + k - Mv^2)(k - mv^2) - k^2} = -\frac{F}{K - Mv^2} \quad \text{(phase requires } -\text{ve sign)}$$

Multiplying out and putting $\mu = m/M$ gives

$$2\left(\frac{v}{\omega}\right)^4 - (4 + \mu)\left(\frac{v}{\omega}\right)^2 + 2 = 0.$$

Since

$$\mu = \frac{600}{3000} = 0.2,$$

$$\left(\frac{v}{\omega}\right)^2 = \frac{4 + \mu}{4} \pm \tfrac{1}{4}\sqrt{(\mu^2 + 8\mu)} = 1.05 \pm 0.32.$$

Thus

$$\frac{v}{\omega} = 1.17 \text{ or } 0.855, \quad \text{and}$$

$$f = 102 \text{ Hz or } 140 \text{ Hz, where } v = 2\pi f.$$

Thus the required frequency range is 102–140 Hz.

A convenient analysis of a system with a vibration absorber can be carried out by using the receptance technique.

Consider the undamped dynamic vibration absorber shown in Fig. 5.18. The system is split into subsystems A and B, where B represents the absorber.

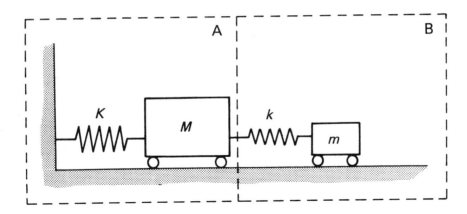

Fig. 5.18.   Subsystem analysis.

For subsystem A (the structure),

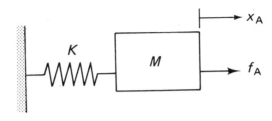

$$f_A = M\ddot{x}_A + Kx_A$$

and

$$\alpha = \frac{1}{K - Mv^2}.$$

For subsystem B (the absorber),

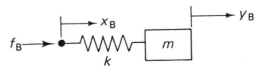

$$f_B = k(x_B - y_B) = m\ddot{y}_B = -mv^2Y_B$$

and

$$\beta = -\left(\frac{k - mv^2}{kmv^2}\right)$$

Thus the frequency equation $\alpha + \beta = 0$ gives

$$Mmv^4 - (mK + Mk + mk)v^2 + Kk = 0,$$

as before.

It is often convenient to solve the frequency equation $\alpha + \beta = 0$ or $\alpha = -\beta$ by a graphical method. In the case of the absorber, both α and $-\beta$ can be plotted as a function of v, and the intersections give the natural frequencies Ω_1 and Ω_2 as shown in Fig. 5.19.

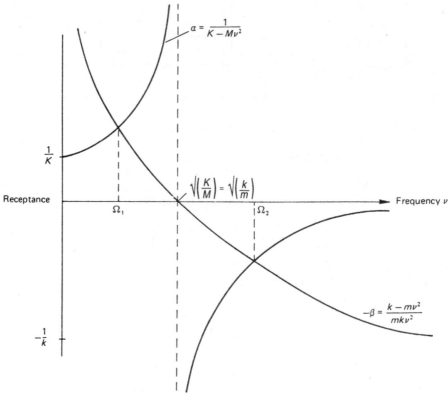

Fig. 5.19.   Subsystem receptance–frequency responses.

This technique is particularly useful when it is required to investigate the effect of several different absorbers, since once the receptance of the primary system is known, it is only necessary to analyse each absorber and not the complete system in each case. Furthermore, sometimes the receptances of structures are measured and are only available in graphical form.

If the proximity of Ω_1 and Ω_2 to ω is likely to be a hazard, damping can be added in parallel with the absorber spring, to limit the response at these frequencies. Unfortunately, if damping is added, the response at the frequency ω will no longer be zero.

The damped dynamic vibration absorber

Fig. 5.20 shows the primary system with a viscous damped absorber added. The equations of motion are

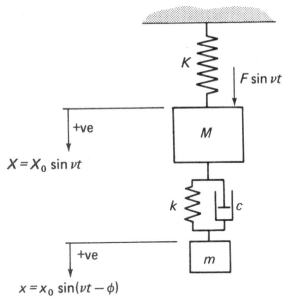

Fig. 5.20.   System with damped vibration absorber.

$$M\ddot{X} = F \sin vt - KX - k(X - x) - c(\dot{X} - \dot{x})$$

and

$$m\ddot{x} = k(X - x) + c(\dot{X} - \dot{x}).$$

Substituting $X = X_0 \sin vt$ and $x = x_0 \sin (vt - \phi)$ gives, after some manipulation,

$$X_0 = \frac{F\sqrt{[(k - mv^2)^2 + (cv)^2]}}{\sqrt{\{[(k - mv^2)(K + k - Mv^2) - k^2]^2 + [(K - Mv^2 - mv^2)cv]^2\}}}$$

It can be seen that when $c = 0$ this expression reduces to that given above for the undamped vibration absorber. Also when c is very large

$$X_0 = \frac{F}{K - (M + m) v^2}$$

For intermediate values of c the primary system response has damped resonance peaks, although the amplitude of vibration does not fall to zero at the original resonance frequency. This is shown in Fig. 5.21.

The response of the primary system can be minimized over a wide range of exciting frequencies by carefully choosing the value of c, and also arranging the system parameters so that the points P_1 and P_2 are at about the same amplitude. However, one of the main advantages of the undamped absorber, that of reducing the vibration amplitude of the primary system to zero at the troublesome resonance, is lost.

A design criterion that has to be carefully considered is the possible fatigue and failure of the absorber spring: this could have severe consequences. In view of this, some damped

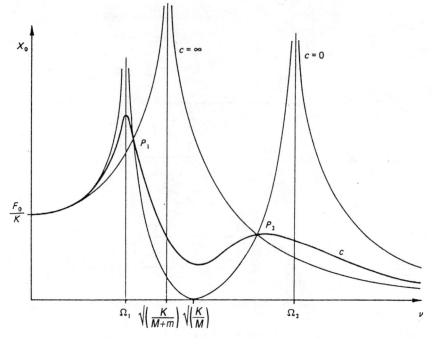

Fig. 5.21.   Effect of absorber damping on system response.

absorber systems dispense with the absorber spring and sacrifice some of the absorber effectiveness. This has particularly wide application in torsional systems, where the device is known as a *Lanchester Damper*.

It can be seen that if $k = 0$,

$$X_0 = \frac{F\sqrt{(m^2 v^4 + c^2 v^2)}}{\sqrt{\{[(K - Mv^2)mv^2]^2 + [(K - Mv^2 - mv^2)cv]^2\}}}.$$

When $c = 0$,

$$X_0 = \frac{F}{K - Mv^2} \quad \text{(no absorber)}$$

and when c is very large,

$$X_0 = \frac{F}{K - (M + m)v^2}.$$

These responses are shown in Fig. 5.22 together with that for the optimum value of c.

The springless vibration absorber is much less effective than the sprung absorber, but has to be used when spring failure is likely, or would prove disastrous.

Vibration absorbers are widely used to control structural resonances. Applications include:

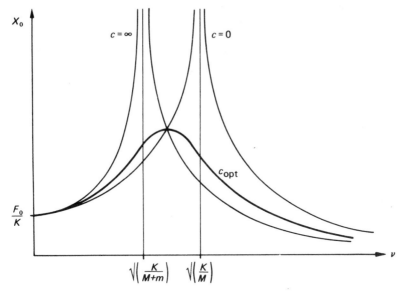

Fig. 5.22.   Effect of Lanchester damper on system response.

1. Machine tools, where large absorber bodies can be attached to the headstock or frame for control of a troublesome resonance.
2. Overhead power transmission lines, where vibration absorbers known as Stockbridge dampers are used for controlling line resonance excited by cross winds.
3. Engine crankshaft torsional vibration, where Lanchester dampers can be attached to the pulley for the control of engine harmonics.
4. Footbridge structures, where pedestrian-excited vibration has been reduced by an order of magnitude by fitting vibration absorbers.
5. Engines, pumps and diesel generator sets where vibration absorbers are fitted so that the vibration transmitted to the supporting structure is reduced or eliminated.

Not all damped absorbers rely on viscous damping; dry friction damping is often used, and the replacement of the spring and damper elements by a single rubber block possessing both properties is fairly common.

A structure or mechanism that has loosely fitting parts is often found to rattle when vibration takes place. Rattling consists of a succession of impacts, these dissipate vibrational energy and therefore rattling increases the structural damping. It is not desirable to have loosely fitting parts in a structure, but an impact damper can be fitted.

An *impact damper* is a hollow container with a loosely fitting body or slug; vibration causes the slug to impact on the container ends, thereby dissipating vibrational energy. The principle of the impact damper is that when two bodies collide some of their energy is converted into heat and sound so that the vibrational energy is reduced. Sometimes the slug is supported by a spring so that advantage can be taken of resonance effects. Careful tuning is required, particularly with regard to slug mass, material and clearance, if the optimum effect is to be achieved. Although cheap and easy to manufacture and install, impact dampers have often been neglected because they are difficult to analyse and design, and their performance can be unpredictable. They are also rather noisy in operation, although the use of PVC impact surfaces can go some way towards reducing this. Some success has been achieved by fitting vibration absorbers with impact dampers. The significant advantage of the impact vibration absorber over the conventional dynamic absorber is the reduction in the amplitude of the primary system both at resonance and at higher frequencies.

5.8   ACTIVE DAMPING SYSTEMS

The damping that occurs in most dynamic systems and structures is passive; that is, once the system has been designed and manufactured the damping element does not change except possibly by ageing. The damping is designed to control the expected excitation and vibration experienced and to keep the dynamic motions and stresses to acceptable levels. However, the damping does not respond to the stimulus in the sense that it adjusts automatically to the required level, so it is considered to be passive. In active damping systems a measure of feedback is provided so that the level of damping is continually adjusted to provide the optimum control of vibration and desired motion levels. This is shown in block diagram form in Fig. 5.23.

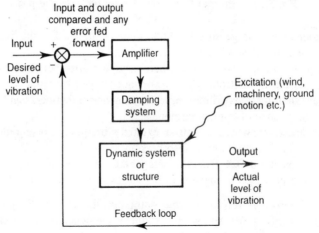

Fig. 5.23.   Block diagram of active damping control system.

The input and output levels of a dynamic quantity such as the motion or vibration are usually measured using a transducer which provides an electrical signal of these responses. The output, that is, the actual vibration level is fed back for comparison with the input and the difference, if any, generates an error signal which is fed to a power amplifier. This amplifier acts on the damping device usually by hydraulic or electrical means to adjust the damping, which acts on the dynamic system or structure. Essentially the control system is an error-actuated power amplifier or servo mechanism which continually adjusts the level of damping so that the structure achieves the desired vibration, motion or stress levels, regardless of the input excitation.

This type of system has been used successfully in active vehicle ride control systems wherein active control is achieved by adjusting the shock absorber or damper settings. In early systems the driver acted as the error detector and feedback device, and moved a knob or lever to adjust the vehicle shock absorbers to give the required damping level according to the desired ride characteristics. These settings were often termed 'sports' and 'normal', 'town' and 'country', 'hard' and 'soft' or simply 'one' and 'two'. Later, microprocessor control was introduced to make the system fully automatic and provide continuous active ride control with optimum damper settings under all conditions. Naturally the active system is more complicated to provide and maintain than a passive system.

Hydraulic shock absorbers or dampers can be adjusted by altering the size of an orifice that the fluid must flow through. This can be done by having a series of holes with a spool valve to divert the flow through one or more of them. The spool in the valve can be moved either hydraulically or by an electrical solenoid device responding to the error-sensing system.

In fabricated structures, joint damping is particularly easy to adapt to active control. In this case the joint clamping force is continually adjusted by hydraulic or electrical means. In this way some of the disadvantages of joint damping such as variations in the coefficient of friction, fretting and partial seizure can be overcome, and the response of a structure optimized whether it be for maximum joint energy dissipation, natural frequency control, level of structural vibration, dynamic stiffness control or noise levels. There is of course some added cost and complexity with active damping systems compared to passive systems. However, there are applications where active damping may be justified. It should also be noted that active damping may be applied to just one or two joints in a structure, or it may be an added damping element acting to reinforce the passive damping which is always present to some extent.

5.9   ENERGY DISSIPATION IN NON-LINEAR STRUCTURES

Although linear analyses explain much of the observed behaviour of vibrating structures, real structures always possess some degree of non-linearity.

A structure is said to be non-linear if the relationship between the excitation and response are not directly proportional. The mass and inertia of a real structure is almost always linear, but both stiffness and damping are always non-linear to some extent although the non-linear effects can be small, particularly in the case of the stiffness. In many cases, non-linearity is localized so that only parts of a structure are non-linear; examples of non-linear structural parts are joints, locally flexible plates, composite materials and buckling struts.

The effects of non-linear springs on the vibration of a structure have already been discussed in section 2.1.3. Analysis techniques may linearize the behaviour over a restricted range, and in practice a hardening spring can lead to instabilities because as a resonance is approached and the amplitude of vibration builds up, the spring stiffness increases, which in turn raises the natural frequency. Further increases in the exciting frequency eventually lead to a resonance jump and the structure settles to a vibration of reduced amplitude and frequency commensurate with the stiffness. A similar instability is experienced when the excitation frequency is decreased through a resonance.

Non-linear damping has been discussed in sections 2.2.2., 2.3.3. and 3.1.5. Analysis techniques often seek to linearize either the damping action or the equations of motion. For example, to linearize the behaviour of a non-linear friction joint, the frequency response function at frequencies close to a resonance of a structure can be found and the parameters linearized over short frequency intervals. This often leads to difficulties when assessing the energy dissipation in non-linear structures.

Because non-linear spring and damping effects are often inseparable, it is generally helpful to consider the energy dissipation in non-linear structures rather than the damping alone. This avoids confusion with the interpretation of the cause of the non-linearities in question, which may be coupled in the non-linear component.

One of the most common sources of non-linearity in structures originates in the joints. Relative interfacial motion may occur on a micro- or macro-slip scale depending upon clamping forces and surface conditions, which results in significant energy dissipation. In addition the clamping force controls the joint stiffness.

Consider the energy that can be dissipated by a simple oscillating joint with metal to metal sliding contact as shown in Fig. 5.24. An exciting force $F \sin vt$ is applied to the joint member of mass m supported by a spring of stiffness k. The other joint member is rigidly fixed. A constant force N is applied normal to the joint interface, where the coefficient of friction is μ.

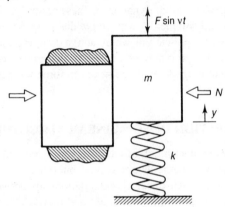

Fig. 5.24.   Metal to metal sliding contact joint.

There is no movement until $F \sin vt \geq \mu N$, and then

$$F \sin vt - \mu N - ky = m\ddot{y}.$$

Since the exciting force is sinusoidal, it is reasonable to assume

$$y = Y \sin (vt - \phi).$$

Hence

$$F \sin vt - \mu N = ky - mv^2 y$$

$$= ky\left(1 - \left(\frac{v}{\omega}\right)^2\right),$$

where

$$\omega = \sqrt{(k/m)},$$

that is,

$$y = \frac{F \sin vt - \mu N}{k\left(1 - \left(\dfrac{v}{\omega}\right)^2\right)}$$

and

$$Y = \frac{F - \mu N}{k\left(1 - \left(\dfrac{v}{\omega}\right)^2\right)}.$$

The energy dissipated per cycle, E, is $4Y\mu N$, so that

$$E = 4\mu N \left[\frac{F - \mu N}{k\left(1 - \left(\dfrac{v}{\omega}\right)^2\right)}\right].$$

For maximum E, $dE/d(\mu N) = 0$, and hence it is found that for E_{max},

$$(\mu N)_{E_{max}} = \frac{F}{2},$$

$$E_{max} = \frac{F^2}{k\left(1 - \left(\dfrac{v}{\omega}\right)^2\right)},$$

and

$$Y_{E_{max}} = \frac{F - F/2}{k\left(1 - \left(\dfrac{v}{\omega}\right)^2\right)} = \tfrac{1}{2}Y_{N = 0};$$

that is, the value of μN for maximum energy dissipation is that value of μN which reduces the damped amplitude to one half of the undamped amplitude.

The ratio $\mu N{:}F$ is important, and by calculating E for various values of $\mu N/F$ the curve shown in Fig. 5.25 is obtained. An optimum value of $\mu N/F$ is seen to exist when $E = E_{max}$; it can also be seen that $E \geq 0.5\, E_{max}$ if $\mu N/F$ is maintained between 0.15 and 0.85, and $E \geq 0.75\, E_{max}$ if $\mu N/F$ is between 0.25 and 0.75.

Since the amplitude of slip under maximum energy dissipation conditions is one half of the amplitude for zero clamping force, this provides a simple practical method for adjusting such a joint to provide maximum energy dissipation. However, the resulting contact pressures are usually too low to be found in structurally necessary joints so that a special type of joint may be required such as that shown in Fig. 5.26. This joint has good load-bearing properties combined with high energy dissipation from both the rubber material and the joint contact mechanism.

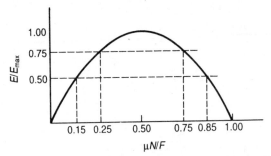

Fig. 5.25.   Effect of $\mu N/F$ on energy dissipated per cycle.

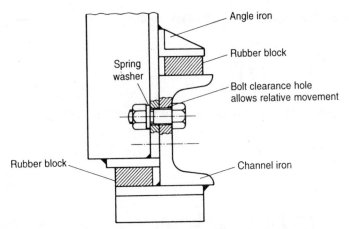

Fig. 5.26.   Joint designed to carry structural load and dissipate vibrational energy by material damping in rubber blocks and controlled relative slip at joint interface. Normal joint clamping force adjusted with bolt and spring washer arrangement.

The effect of adding a friction joint to a beam type structure can be analysed by considering the cantilever shown in Fig. 5.27. An exciting force $F \sin vt$ is applied a distance a from the root.

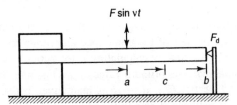

Fig. 5.27.   Cantilever beam with friction joint.

If the joint frictional force F_d can be represented by a series of linear periodic functions then

$$y_a = \alpha_{aa} F \sin vt + \alpha_{ab} F_d,$$

$$y_b = \alpha_{ba} F \sin vt + \alpha_{bb} F_d$$

and

$$y_c = \alpha_{ca} F \sin vt + \alpha_{cb} F_d,$$

where α_{aa} ... are receptances in series form. If it is assumed that F_d is always out of phase with F and that it contains a dominant harmonic component at frequency v, then

$$y_b = (\alpha_{ba} F - \alpha_{bb} F_d) \sin vt.$$

The energy dissipated per $\frac{1}{4}$ cycle is

$$\int_0^{\pi/2v} F_d \sin vt \cdot \dot{y}_b \, dt$$

which is also equal to $\mu N Y_b$.

Carrying out the integration and substituting

$$Y_b = \alpha_{ba} F - \alpha_{bb} F_d$$

gives

$$F_d = 2\mu N;$$

that is, this technique for linearization of the damping replaces the actual frictional force during slipping, μN, by a sinusoidally varying force of amplitude $2\mu N$ directly out of phase with the excitation.

The effect of changes in the joint clamping force N on the energy dissipation capabilities of the joint are readily found.

Many structures rely on the ability of the joints to transmit translational forces through shear in the bolts to achieve their static stiffness. Bolts are normally tightened as much as possible to maximize the bending moments which can be transmitted through a joint and thereby increase the static stiffness. This type of joint can give good frictional energy dissipation with translational slip if the translational forces are transmitted only by friction throughout the joint, but in many structures such as frameworks, this form of slip is not practicable. However, it is possible for bolt tightening to be controlled to allow joints to slip in rotation and provide significant energy dissipation.

Consider the general structural joint shown in Fig. 5.28. Excitation is provided by the harmonic force $F \sin vt$, and the friction torque by T_d which acts on a representative joint.

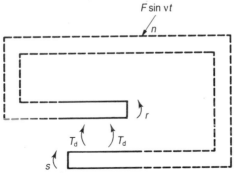

Fig. 5.28.   Structural joint with friction torque T_d.

If T_d is assumed to be harmonic and to lag the relative slip velocity by 180° if slipping occurs, the displacement response at a general coordinate i is

$$X_i = \alpha_{in} F \sin vt + (\alpha_{ir} - \alpha_{is}) T_d \sin vt,$$

and the relative rotation across the joint is

$$\phi = X_r - X_s = (\alpha_{rn} - \alpha_{sn}) F \sin vt + (\alpha_{rr} + \alpha_{ss} - \alpha_{rs} - \alpha_{sr}) T_d \sin vt.$$

The value of the limiting friction at the joint interfaces, the applied forces and the frequency dictate whether slipping occurs, which gives three distinct response regimes for ϕ.

(i) T_d effectively zero (free joint).

$$(X_i)_{T_d = 0} = \alpha_{in} F \sin vt.$$

(ii) $T_d = T_L$, the torque required to lock the joint and prohibit slip.

$$(X_i)_{T_d = T_L} = (X_i)_{T_d = 0} + (\alpha_{ir} - \alpha_{is})T_L$$

(iii) Slip response, $T_L > T_d > 0$

$$(X_i)_{T_d} = \left\{ (X_i)_{T_d = 0} + \frac{T_d}{T_L} \left[(X_i)_{T_d = T_L} - (X_i)_{T_d = 0} \right] \cos \theta \right\} \sin vt$$

$$+ \left\{ \frac{T_d}{T_L} \left[(X_i)_{T_d = T_L} - (X_i)_{T_d = 0} \right] \sin \theta \right\} \cos vt$$

where $\cos \theta = T_d/T_L$.

These expressions enable the energy dissipation by relative rotational slip in a joint to be evaluated for any clamping torque.

The effect of friction damping on the vibration of plates can be considered by investigating laminated plates clamped together to generate interfacial in-plane friction forces. When the plate vibrates the laminates experience slight relative interfacial slip. High energy dissipation can be achieved but there is an associated loss in static stiffness compared with a solid plate, which may be important in some applications; that is, such a plate possesses both non-linear stiffness and damping effects which are inseparable.

6

Problems

6.1 THE VIBRATION OF STRUCTURES WITH ONE DEGREE OF FREEDOM

1. A structure is modelled by a rigid horizontal member of mass 3000 kg, supported at each end by a light elastic vertical member of flexural stiffness 2 MN/m.
    Find the frequency of small-amplitude horizontal vibrations of the rigid member.
2. Part of a structure is modelled by a thin rigid rod of mass m pivoted at the lower end, and held in the vertical position by two springs, each of stiffness k, as shown.
    Find the frequency of small-amplitude oscillation of the rod about the pivot.

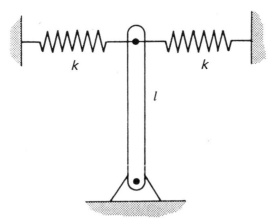

3. A uniform beam of length 8 m, simply supported at the ends, carries a uniformly distributed mass of 300 kg/m and three bodies, one of mass 1000 kg at mid-span and

two of mass 1500 kg each, at 2 m from each end. The second moment of area of the beam is 10^{-4} m^4 and the modulus of elasticity of the material is 200 GN/m^2.

Estimate the lowest natural frequency of flexural vibration of the beam assuming that the deflection y_x at a distance x from one end is given by:

$$y_x = y_c \sin \pi(x/l),$$

where y_c is the deflection at mid-span and l is the length of the beam.

4. A section of steel pipe in a distillation plant is 80 mm in diameter, 5 mm thick and 4 m long. The pipe may be assumed to be built-in at each end, so that the deflection y, at a distance x from one end of a pipe of length l is

$$y = \frac{mg}{24\ EI}\ x^2\ (l - x)^2,$$

m being the mass per unit length.

Calculate the lowest natural frequency of transverse vibration of the pipe when full of water. Take $\rho_{\text{steel}} = 7750$ kg/m^3, $\rho_{\text{water}} = 930$ kg/m^3 and $E_{\text{steel}} = 200$ GN/m^2.

5. A uniform horizontal steel beam is built in to a rigid structure at one end and pinned at the other end; the pinned end cannot move vertically but is otherwise unconstrained. The beam is 8 m long, the relevant flexural second moment of area of the cross section is 4.3×10^6 mm^4, and the beam's own mass together with the mass attached to the beam is equivalent to a uniformly distributed mass of 600 kg/m.

Using a combination of sinusoidal functions for the deflected shape of the beam, estimate the lowest natural frequency of flexural vibrations in the vertical plane.

6. Estimate the lowest frequency of natural transverse vibration of a chimney 100 m high, which can be represented by a series of lumped masses M at distances y from its base as follows:

y (m)	20	40	60	80	100
M (10^3 kg)	700	540	400	280	180

With the chimney considered as a cantilever on its side the static deflection in bending, x along the chimney is calculated to be

$$x = X\left(1 - \cos \pi \frac{y}{2l}\right),$$

where $l = 100$ m, and $X = 0.2$ m.

How would you expect the actual frequency to compare with the frequency that you have calculated?

7. Estimate the lowest natural frequency of a light beam 7 m long carrying six concentrated masses equally spaced along its length. The measured static deflections under each mass are:

Mass (kg)	1070	970	370	370	670	670
Deflection (mm)	2.5	2.8	5.5	5.0	2.5	1.0

8. An elastic part of a structure has a dynamic force-deflection loop at a frequency of 10 rad/s as shown below.

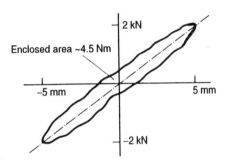

Find
(a) the stiffness k,
(b) assuming viscous damping find c and ζ, and
(c) assuming hysteretic damping find η.

9. A solid steel shaft, 25 mm in diameter and 0.45 m long, is mounted in long bearings in a rigid frame at one end and has at its other end, which is unsupported, a steel flywheel. The flywheel can be treated as a rim 0.6 m in outer diameter and 20 mm square cross section, with rigid spokes of negligible mass lying in the mid-plane of the rim.
    Find the frequency of free flexural vibrations.

10. A uniform rigid building, height 30 m and cross section 10 m × 10 m, rests on an elastic soil of stiffness 0.6×10^6 N/m^3. (Stiffness is defined as the force per unit area to produce unit deflection.)
    If the mass of the building is 2×10^6 kg and its inertia about its axis of rocking at the base is 500×10^6 kg m^2, calculate the period of the rocking motion (small amplitudes).
    What wind speed would excite this motion if the Strouhal number is 0.22? Calculate also the maximum height the building could be before becoming unstable.

11. A single degree of freedom system with a body of mass 10 kg, a spring of stiffness 1 kN/m and negligible damping is subjected to an input force F which varies with time as shown overleaf.

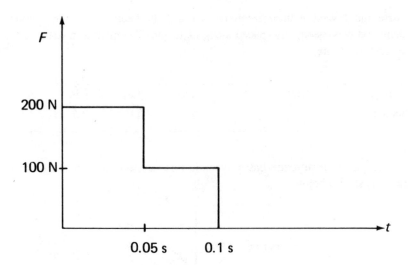

Determine the amplitude of free vibration of the body after the force is removed.

12. A uniform rigid tower of height 30 m and cross section 3 m × 3 m, is symmetrically mounted on a rigid foundation of depth 2 m and section 5 m × 5 m. The mass of the tower is calculated to be 1.5×10^5 kg, and of the foundation, 10^3 kg. The foundation rests on an elastic soil which has a uniform stiffness of 2×10^6 N/m³. (Stiffness is defined as the force per unit area to produce unit deflection.)

   If the mass moment of inertia of the tower and foundation about its axis of rocking at the base of the foundation is 6×10^7 kg m², find the period of small-amplitude rocking motion. The axis of rocking is parallel to a side of the foundation.

   What is the greatest height the tower could have and still be stable on this foundation?

13. The foundation of a rigid tower is a circular concrete block of diameter D, set into an elastic soil. The effective stiffness of the soil, k, is defined as the force per unit area to produce a unit displacement and is constant for small deflections. The tower is uniform with a total mass M. The centre of mass is situated on the centre line at a height h above the base. The moment of inertia of the tower about an axis of rocking at the base is I.

   Show that the natural frequency of rocking is given by:

$$\frac{1}{2\pi} \sqrt{\left(\frac{\pi k D^4/64 - Mgh}{I} \right)} \ \text{Hz.}$$

14. A body supported by an elastic structure performs a damped oscillation of period 1 s, in a medium whose resistance is proportional to the velocity. At a given instant the amplitude was observed to be 100 mm, and in 10 s this had reduced to 1 mm.

   What would be the period of the free vibration if the resistance of the medium were negligible?

15. To determine the amount of damping in a bridge it was set into vibration in the fundamental mode by dropping a weight on it at centre span. The observed frequency was 1.5 Hz, and the amplitude was found to have decreased to 0.9 of the initial maximum after 2 s. The equivalent mass of the bridge (estimated by the Rayleigh Energy method) was 10^5 kg.

    Assuming viscous damping and simple harmonic motion, calculate the damping coefficient, the logarithmic decrement and the damping ratio.

16. A new concert hall is to be protected from the ground vibrations from an adjacent highway by mounting the hall on rubber blocks. The predominant frequency of the sinusoidal ground vibrations is 40 Hz, and a motion transmissibility of 0.1 is to be attained at that frequency.

    Calculate the static deflection required in the rubber blocks, assuming that these act as linear, undamped springs.

17. When considering the vibrations of a structure, what is meant by the Q factor? Derive a simple relationship between the Q factor and the damping ratio for a single degree of freedom system with light viscous damping.

    Measurements of the vibration of a bridge section resulting from impact tests show that the period of each cycle is 0.6 s, and that the amplitude of the third cycle is twice the amplitude of the ninth cycle. Assuming the damping to be viscous estimate the Q factor of the section.

    When a vehicle of mass 4000 kg is positioned at the centre of the section the period of each cycle increases to 0.62 s; no change is recorded in the rate of decay of the vibration. What is the effective mass of the section?

18. The vibration on the floor of a laboratory is found to be simple harmonic motion at a frequency in the range 15–60 Hz, (depending on the speed of some nearby reciprocating plant). It is desired to install in the laboratory a sensitive instrument which requires insulating from the floor vibration. The instrument is to be mounted on a small platform which is supported by three similar springs resting on the floor, arranged to carry equal loads; the motion is restrained to occur in a vertical direction only. The combined mass of the instrument and the platform is 40 kg: the mass of the springs can be neglected and the equivalent viscous damping ratio of the suspension is 0.2.

    Calculate the maximum value for the spring stiffness, if the amplitude of the transmitted vibration is to be less than 10% of that of the floor vibration over the given frequency range.

19. A machine of mass 520 kg produces a vertical disturbing force which oscillates sinusoidally at a frequency of 25 Hz. The force transmitted to the floor is to have an amplitude, at this frequency, not more than 0.4 times that of the disturbing force in the machine, and the static deflection of the machine on its mountings is to be as small as possible consistent with this.

    For this purpose, rubber mountings are to be used, which are available as units, each of which has a stiffness of 359 kN/m and a damping coefficient of 2410 N s/m. How many of these units are needed?

20. The basic element of many vibration-measuring devices is the seismic unit, which consists of a mass m supported from a frame by a spring of stiffness k in parallel with a damper of viscous damping coefficient c. The frame of the unit is attached to the structure whose vibration is to be determined, the quantity measured being z, the relative motion between the seismic mass and the frame. The motion of both the structure and the seismic mass is translation in the vertical direction only.

   Derive the equation of motion of the seismic mass, assuming that the structure has simple harmonic motion of circular frequency v, and deduce the steady state amplitude of z.

   Given that the undamped natural circular frequency ω of the unit is much greater than v, show why the unit may be used to measure the acceleration of the structure.

   Explain why in practice some damping is desirable.

   If the sensitivity of the unit (that is, the amplitude of z as a multiple of the amplitude of the acceleration of the frame) is to have the same value when $v = 0.2\,\omega$ as when $v \ll \omega$, find the necessary value of the damping ratio.

21. A two-wheeled trailer of sprung mass 700 kg is towed at 60 km/h, along an undulating straight road whose surface may be considered sinusoidal. The distance from peak to peak of the road surface is 30 m, and the height from hollow to crest is 0.1 m. The effective stiffness of the trailer suspension is 60 kN/m, and the shock absorbers, which provide linear viscous damping, are set to give a damping ratio of 0.67.

   Assuming that only vertical motion of the trailer is excited, find the absolute amplitude of this motion and its phase angle relative to the undulations.

22. Find the Fourier series representation of the following triangular wave.

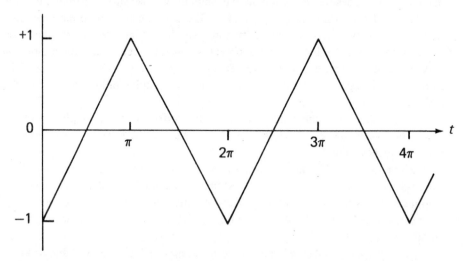

23. A wooden floor, 6 m by 3 m, is simply supported along the two shorter edges. The mass is 300 kg and the static deflection at the centre under the self-weight is 7 mm.

It is proposed to determine the dynamic properties of the floor by dropping a sand bag of 50 kg mass on it at the centre, and to measure the response at that position with an accelerometer and a recorder.

In order to select the instruments required, estimate:

(i) the frequency of the fundamental mode of vibration that would be recorded,

(ii) the number of oscillations at the fundamental frequency for the signal amplitude on the recorder to be reduced to half, assuming a loss factor of 0.05, and

(iii) the height of drop of the sand bag so that the dynamic deflection shall not exceed 10 mm, and the corresponding maximum acceleration.

24. The figure shows a diagrammatic end view of one half of a swing-axle suspension of a motor vehicle which consists of a horizontal half-axle OA pivoted to the chassis at O, a wheel rotating about the centre line of the axle and a spring of stiffness k and a viscous damper with a damping coefficient c both located vertically between the axle and the chassis. The mass of the half-axle is m_1 and its radius of gyration about O is h. The mass of the wheel is m_2 and it may be regarded as a thin uniform disc having an external radius r and located at a horizontal distance s from the pivot O. The spring and damper are located at horizontal distances q and p from the pivot O, as shown.

Derive the equation for angular displacement of the axle-wheel assembly about the pivot O, and obtain from it an expression for the frequency of damped free oscillations of the assembly. Express this frequency in terms of the given parameters and the undamped natural frequency of the assembly.

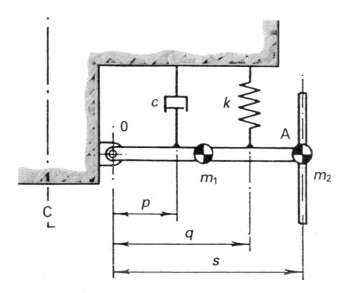

25. The T-shaped body shown overleaf pivots about a point O on a horizontal ground surface and is held upright, so that its mass centre G is a distance h vertically above O, by two springs pinned to it and to the ground. Each spring has a stiffness k and its vertical

centre line is at a distance c from the pivot O. The T-shaped body has a mass m and a radius of gyration r about its mass centre.

The body is acted on by a force F whose line of action is horizontal and at a height d above the ground, where $d > h$. Derive an expression for the rotation of the body if the force rises suddenly from zero to F, assuming that the angular displacement of the body is small.

If the suddenly applied force F drops equally suddenly to zero after a time t_0 from its original application, derive the equation of the rotational motion of the body for times greater than t_0.

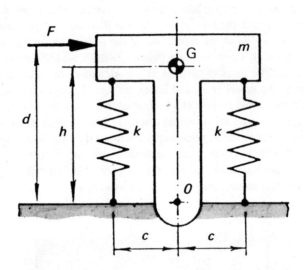

26. The figure shows a vibration exciter which consists of two contra-rotating wheels, each carrying an eccentric body of mass 0.1 kg at an effective radius of 10 mm from its axis of rotation. The vertical positions of the eccentric bodies occur at the same instant. The total mass of the exciter is 2 kg and damping is negligible.

Find a value for the stiffness of the spring mounting so that a force of amplitude 100 N, due to rotor unbalance, is transmitted to the fixed support when the wheels rotate at 150 rad/s.

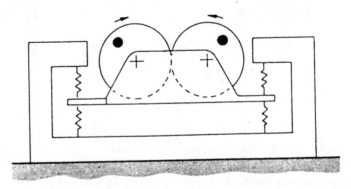

6.2 THE VIBRATION OF STRUCTURES WITH MORE THAN ONE DEGREE OF FREEDOM

27. A two-storey building is represented by the two degree of freedom lumped mass system shown below.

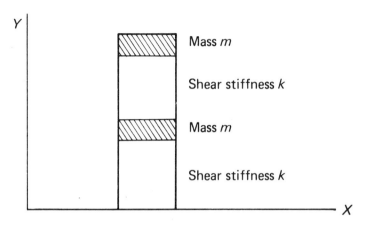

Obtain the frequency equation for swaying motion in the X–Y plane; hence calculate the natural frequencies and sketch the corresponding mode shapes.

28. A vehicle has a mass of 2000 kg and a 3 m wheelbase. The mass moment of inertia about the centre of mass is 500 kg m^2, and the centre of mass is located 1 m from the front axle. Considering the vehicle as a two degree of freedom system, find the natural frequencies and the corresponding modes of vibration, if the front and rear springs have stiffnesses of 50 kN/m and 80 kN/m, respectively.

The expansion joints of a concrete road are 5 m apart. These joints cause a series of impulses at equal intervals to vehicles travelling at a constant speed. Determine the speeds at which pitching motion and up and down motion are most likely to arise for the above vehicle.

29. To analyse the vibrations of a two-storey building it is represented by the lumped mass system shown overleaf, where $m_1 = \frac{1}{2} m_2$, and $k_1 = \frac{1}{2} k_2$ (k_1 and k_2 represent the shear stiffnesses of the parts of the building shown).

Calculate the natural frequencies of free vibrations, and sketch the corresponding mode shapes of the building, showing the amplitude ratios.

If a horizontal harmonic force $F_1 \sin vt$ is applied to the top floor, determine expressions for the amplitudes of the steady state vibration of each floor.

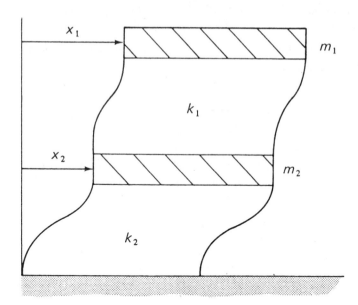

30. The rigid beam, shown in its position of static equilibrium in the figure, has a mass m
and a mass moment of inertia $2\ ma^2$ about an axis perpendicular to the plane of the
diagram, and through its centre of mass G.

    Assuming no horizontal motion of G, find the frequencies of small oscillations in
the plane of the diagram, and the corresponding positions of the nodes.

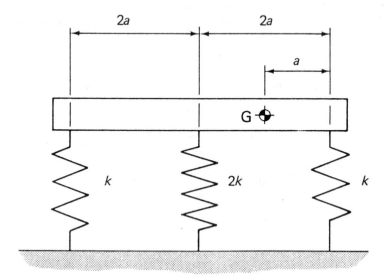

31. A small electronic package is supported by springs as shown opposite. The mass of the
package is m, each spring has a constant axial stiffness k, and damping is negligible.

Considering motion in the plane of the figure only, and assuming that the amplitude of vibration of the package is small enough for the lateral spring forces to be negligible, write down the equations of motion and hence obtain the frequencies of free vibration of the package.

Explain how the vibration mode shapes can be found.

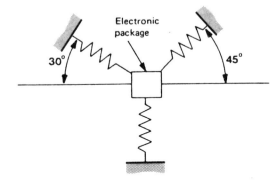

32. Explain, in one sentence each, what is meant by a *natural frequency* and *mode shape* of a dynamic system.

33. Part of a machine can be modelled by the system shown. Two uniform discs A and B, which are free to rotate about fixed parallel axes through their centres, are coupled by a spring. Similar springs connect the discs to the fixed frame as shown in the figure. Each of the springs has a stiffness of 2.5 kN/m which is the same in tension or compression. Disc A has a mass moment of inertia about its axis of rotation of 0.05 kg m^2, and a radius of 0.1 m, whilst for the disc B the corresponding figures are 0.3 kg m^2 and 0.2 m. Damping is negligible.

Determine the natural frequencies of small amplitude oscillation of the system and the corresponding mode shapes.

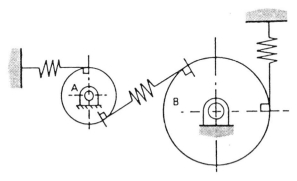

34. Part of a building structure is modelled by the triple pendulum shown overleaf. Obtain the equations of motion of small-amplitude oscillation in the plane of the figure by using the Lagrange equation.

Hence determine the natural frequencies of the structure and the corresponding mode shapes.

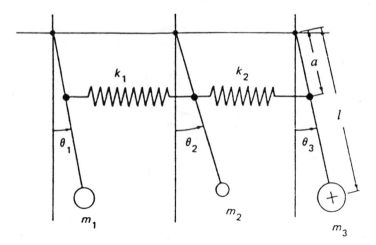

35. A simplified model for studying the dynamics of a motor vehicle is shown. The body
has a mass M, and a moment of inertia about an axis through its mass centre of I_G. It
is considered to be free to move in two directions – vertical translation and rotation in
the vertical plane. Each of the unsprung wheel masses, m, are free to move in vertical
translation only. Damping effects are ignored.

   (i) Derive equations of motion for this system. Define carefully the coordinates
   used.
   (ii) Is it possible to determine the natural frequency of a 'wheel hop' mode without
   solving all the equations of motion? If not, suggest an approximation that might
   be made, in order to obtain an estimate of the wheel hop frequency, and calculate
   such an estimate given the following data:

$$k = 20 \text{ kN/m}; \ K = 70 \text{ kN/m}; \ m = 22.5 \text{ kg}.$$

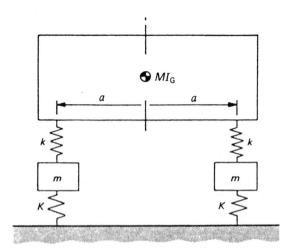

36. If the building in problem 27 were enlarged by adding a further floor of mass m and shear stiffness k on top of the existing building, obtain the frequency equation for the three degree of freedom system formed. Given m and k contemplate how this equation may be solved. What if the building were 20 storeys high?

37. To analyse the vibration of a two-coach rail unit, it is modelled as the system shown. Each coach is represented by a rigid uniform beam of length l and mass m; the coupling is a simple ball-joint. The suspension is considered to be three similar springs, each of stiffness k, positioned as shown. Damping can be neglected.

Considering motion in the plane of the figure only, obtain the equations of motion for small-amplitude free vibrations, and hence obtain the natural frequencies of the system.

Explain how the mode shapes may be found.

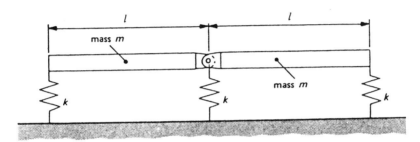

38. A bridge structure is modelled by a simply supported beam of length l, with three equal bodies each of mass m attached to it at equal distances as shown. Show that the influence coefficients are (where $\Delta = l^3/256 EI$):

$$\alpha_{11} = 3\Delta, \qquad \alpha_{12} = 3.67\Delta, \qquad \alpha_{13} = 2.33\Delta,$$
$$\alpha_{21} = 3.67\Delta, \qquad \alpha_{22} = 5.33\Delta, \qquad \alpha_{23} = 3.67\Delta,$$
$$\alpha_{31} = 2.33\Delta, \qquad \alpha_{32} = 3.67\Delta, \qquad \alpha_{33} = 3\Delta.$$

Proceed to find the flexibility matrix and, by iteration, deduce the lowest natural frequency and associated mode shape.

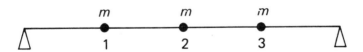

39. A solid cylinder, has a mass M and radius R. Pinned to the axis of the cylinder is an arm of length l which carries a bob of mass m as shown overleaf. Obtain the natural frequency of free vibration of the bob.

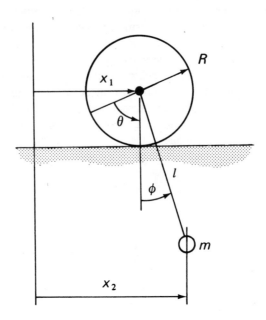

40. An aeroplane has a fuselage mass of 4000 kg. Each wing has an engine of mass 500 kg, and a fuel tank of mass 200 kg at its tip, as shown. Neglecting the mass of each wing, calculate the frequencies of flexural vibrations in a vertical plane. Take the stiffness of the wing sections to be $3k$ and k as shown, where $k = 100$ kN/m.

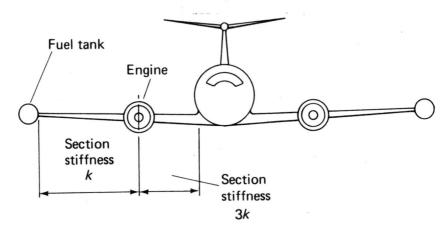

41. A marine propulsion installation is shown in the figure opposite. For the analysis of torsional vibration, the installation can be modelled as the system shown, where the mass moments of inertia for the engine, gearbox and propeller taken about the axis of rotation are I_E, I_G and I_P respectively, and the stiffnesses of the gearbox and propeller shafts are k_G and k_P respectively. The numerical values are

$I_E = 0.8$ kg m^2,
$I_G = 0.3$ kg m^2,
$I_P = 2.0$ kg m^2,
$k_G = 400$ kNm/rad,
$k_P = 120$ kNm/rad,

and damping can be neglected.

Calculate the natural frequencies of free torsional vibration and give the positions of the node for each frequency.

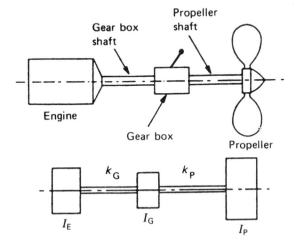

42. A machine is modelled by the system shown. The masses of the main elements are m_1 and m_2, and the spring stiffnesses are as shown. Each roller has a mass m, diameter d, and mass moment of inertia J about its axis, and rolls without slipping.

Considering motion in the longitudinal direction only, use Lagrange's equation to obtain the equations of motion for small free oscillations of the system. If $m_1 = 4m$, $m_2 = 2m$ and $J = md^2/8$, deduce the natural frequencies of the system and the corresponding mode shapes.

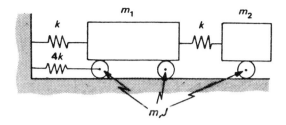

43. Vibrations of a particular structure can be analysed by considering the equivalent system shown overleaf.

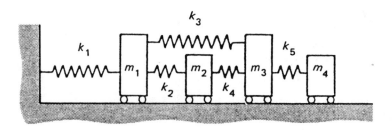

The bodies are mounted on small frictionless rollers whose mass is negligible, and motion occurs in a horizontal direction only.

Write down the equations of motion of the system and determine the frequency equation in determinant form. Indicate how you would

(i) solve the frequency equation, and
(ii) determine the mode shapes associated with each natural frequency.

Briefly describe how the Lagrange equation could be used to obtain the natural frequencies of free vibration of the given system.

44. A simply supported beam of negligible mass and length l, has three bodies each of mass m attached as shown. The influence coefficients are, using standard notation,

$$\alpha_{11} = 3l^3/256\ EI, \quad \alpha_{31} = 2.33l^3/256\ EI,$$
$$\alpha_{21} = 3.67l^3/256\ EI, \quad \alpha_{22} = 5.33l^3/256\ EI.$$

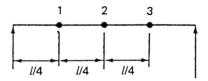

Write down the flexibility matrix, and determine by iteration the frequency of the first mode of vibration, correct to 2 significant figures, if $EI = 10\ \text{Nm}^2$, $m = 2$ kg and $l = 1$ m.

Comment on the physical meaning of the eigenvector you have obtained, and use the orthogonality principle to obtain the frequencies of the higher modes.

45. A structure is modelled by three identical long beams and rigid bodies, connected by two springs as shown opposite. The rigid bodies are each of mass M and the mass of the beams is negligible. Each beam has a transverse stiffness K at its unsupported end; and the springs have stiffnesses k and $2k$ as shown.

Determine the frequencies and corresponding mode shapes of small-amplitude oscillation of the bodies in the plane of the figure. Damping can be neglected.

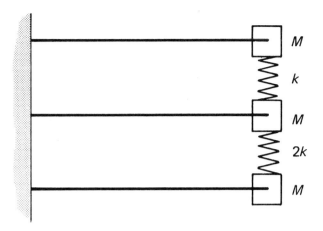

46. Find the dynamic matrix of the system shown.
    If $k = 20$ kN/m and $m = 5$ kg, find the lowest natural frequency of the system and the associated mode shape.

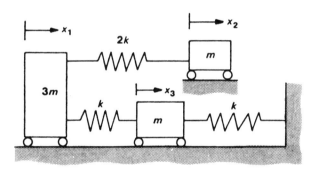

47. A structure is modelled by the three degree of freedom system shown overleaf. Only translational motion in a vertical direction can occur.

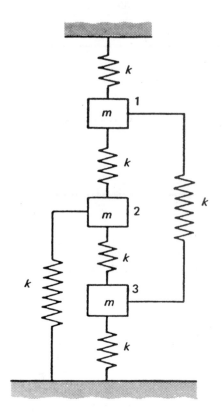

Show that the influence coefficients are

$$\alpha_{11} = \alpha_{22} = \alpha_{33} = \tfrac{1}{2}k$$

and

$$\alpha_{21} = \alpha_{31} = \alpha_{32} = \tfrac{1}{4}k,$$

and proceed to find the flexibility matrix. Hence obtain the lowest natural frequency of the system and the corresponding mode shape.

48. A delicate instrument is to be mounted on an antivibration installation so as to minimize the risk of interference caused by groundborne vibration. An elevation of the installation is shown opposite, and the point A indicates the location of the most sensitive part of the instrument. The installation is free to move in the vertical plane, but horizontal translation is not to be considered.

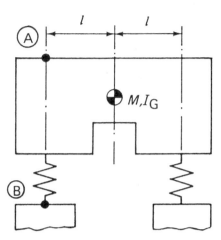

It is decided to use as a design criterion the transmissibility T_{AB}, being the sinusoidal vibration amplitude at A for a unit amplitude of vibration on the ground at B. One of the major sources of groundborne vibration is a nearby workshop where there are several machines which run at 3000 rev/min. Accordingly, it is proposed that the installation should have a transmissibility $|T_{AB}|$ of 1% at 50 Hz.

Given the following data:

$$M = 3175 \text{ kg}; \; l = 0.75 \text{ m}; \; R = 0.43 \text{ m (where } I_G = MR^2);$$

determine the maximum value of stiffness K that the mounts may possess in order to meet the requirement, and find the two natural frequencies of the installation.

Repeat the analysis using a simpler model of the system having just one degree of freedom – vertical translation of the whole installation – and establish whether this simpler approach provides an acceptable means of designing such a vibration isolation system.

For the purpose of these basic isolation design calculations, damping may be ignored.

49. In a vibration isolation system, a group of machines are firmly mounted together onto a rigid concrete raft which is then isolated from the foundation by four antivibration pads. For the purposes of analysis, the system may be modelled as a symmetrical body of mass 1150 kg and moments of inertia about rolling and pitching axes through the mass centre of 175 kgm² and 250 kgm², respectively, supported at each corner by a spring of stiffness 7.5×10^5 N/m.

The model is shown overleaf.

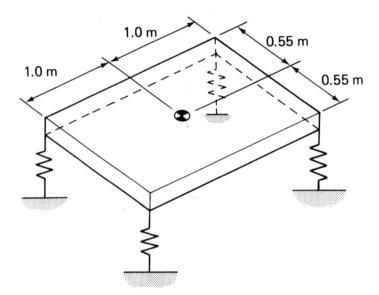

The major disturbing force is generated by a machine at one corner of the raft and may be represented by a harmonically varying vertical force with a frequency of 50 Hz, acting directly through the axis of one of the mounts.

(i) Considering vertical vibration only, show that the force transmitted to the foundation by each mount will be different, and calculate the magnitude of the largest, expressed as a percentage of the excitation force.

(ii) Identify the mode of vibration that is responsible for the largest component of this transmitted force and suggest ways of improving the isolation performance using the same mounts but *without* modifying the raft.

(iii) Show that a considerable improvement in isolation would be obtained by moving the disturbing machine to the centre of the raft, and calculate the transmitted force for this case, again expressed as a percentage of the exciting force.

50. Find the driving point impedance of the system shown. The bodies move on frictionless rollers in a horizontal direction only.

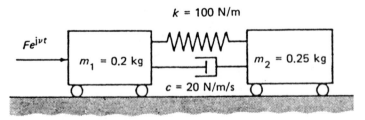

Hence show that the amplitude of vibration of body 1 is

$$\frac{\sqrt{[(72\,000 + 2620v^2 + 0.2v^4)^2 + (20v^3)^2]}}{v^2\,(0.04v^4 + 1224v^2 + 32\,400)}F.$$

51. Find the driving point mobility of the system shown; only motion in the vertical direction occurs and damping is negligible.

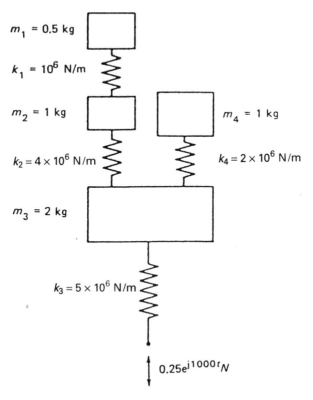

$m_1 = 0.5$ kg

$k_1 = 10^6$ N/m

$m_2 = 1$ kg

$m_4 = 1$ kg

$k_2 = 4 \times 10^6$ N/m

$k_4 = 2 \times 10^6$ N/m

$m_3 = 2$ kg

$k_3 = 5 \times 10^6$ N/m

$0.25e^{j1000t}N$

Hence obtain the frequency equation; check your result by using a different method of analysis.

6.3   THE VIBRATION OF CONTINUOUS STRUCTURES

52. A uniform beam of length l is built-in at one end, and rests on a spring of stiffness k at the other, as shown.

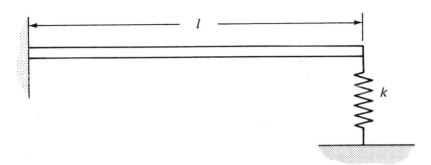

l

k

Determine the frequency equation for small-amplitude transverse vibration, and show how the first natural frequency changes as k increases from zero at the free end, to infinity, at the simply supported end.

Comment on the effect of the value of k on the frequency of the 10th mode.

53. A structure is modelled as a uniform beam of length l, hinged at one end, and resting on a spring of stiffness k at the other, as shown.

Determine the first three natural frequencies of the beam, and sketch the corresponding mode shapes.

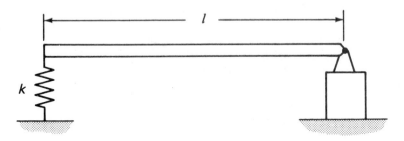

54. Part of a structure is modelled as a uniform cross-section beam having a pinned attachment at one end and a sliding constraint at the other (where it is free to translate, but not to rotate) as shown.

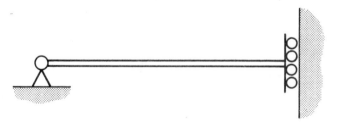

(i) Derive the frequency equation for this beam and find expressions for the nth natural frequency and the corresponding mode shape. Sketch the shapes of the first three modes.

(ii) The beam is to be stiffened by adding a spring of stiffness k to the sliding end. Derive the frequency equation for this case and use the result to deduce the frequency equation for a pinned clamped beam.

(iii) Estimate how much the fundamental frequency of the original beam is raised by adding a very stiff spring to its sliding end.

55. A portal frame consists of three uniform beams, each of length l, mass m, and flexural rigidity EI, attached as shown. There is no relative rotation between beams at their joints.

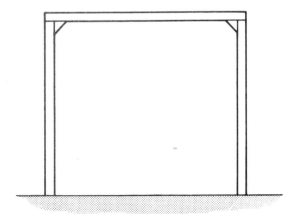

    Show that the fundamental frequency of free vibration, in the plane of the frame is $0.5\sqrt{(EI/ml^3)}$ Hz.

56. A uniform cantilever of length l and flexural rigidity EI, is subjected to a transverse harmonic exciting force $F \sin vt$ at the free end. Show that the displacement at the free end is

$$\left[\frac{\sin \lambda l \cdot \cosh \lambda l - \cos \lambda l \cdot \sinh \lambda l}{EI\lambda^3 (1 + \cos \lambda l \cdot \cosh \lambda l)} \right] F \sin vt,$$

    where $\lambda = (\rho A v^2/EI)^{1/4}$.

57. A thin rectangular plate has its long sides simply supported, and both its short sides unsupported. Find the first three natural frequencies of flexural vibration, and sketch the corresponding mode shapes.

58. Derive the frequency equation for flexural vibration of a uniform beam that is pinned at one end and free at the other.

    Show that the fundamental mode of vibration has a natural frequency of zero, and explain the physical significance of this mode.

    Obtain an approximate value for the natural frequency of the first bending mode of vibration, and compare this with the corresponding value for a beam that is rigidly clamped at one end and free at the other.

6.4 DAMPING IN STRUCTURES

59. The 'half-power' method of determining the damping in a particular mode of vibration from a receptance plot can be extended to a more general form in which the two points used – one below resonance and one above – need not be at an amplitude exactly 0.707 times the peak value.

(i) A typical Nyquist plot of the receptance for a single degree of freedom system
with structural damping is shown, with two points corresponding to frequencies
v_1 and v_2. The natural frequency, ω, is also indicated. Prove that the damping loss
factor, η, is given exactly by:

$$\eta = [(v_2^{\,2} - v_1^{\,2})/\omega^2]\,[(\tan \tfrac{1}{2}\phi_1 + \tan \tfrac{1}{2}\phi_2)]^{-1},$$

where ϕ_1 and ϕ_2 are the angles subtended by points 1 and 2 with the resonance
point and the centre of the circle. Show how this expression relates to the half-
power points formula.

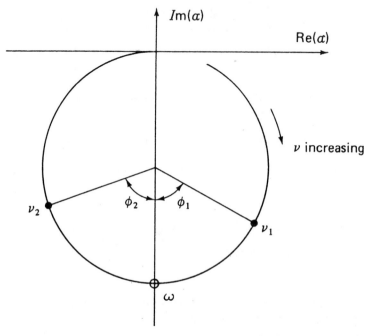

(ii) Some receptance data from measurements on a practical structure are listed in the
table opposite. By application of the formula above to the data given, obtain a
best estimate for the damping of the mode under investigation.

Frequency (Hz)	Receptance	
	Modulus ($\times 10^{-7}$ m/N)	Phase (degrees)
5.86	12.3	24
5.87	12.6	29
5.88	12.5	36.5
5.89	12.0	41
5.90	11.3	57
5.91	10.1	66
5.92	8.8	74
5.93	7.0	78
5.94	5.6	78

60. A large symmetric machine tool structure is supported by four suspension units, one at each corner, intended to provide isolation against vibration. Each unit consists of a primary spring (which can be considered massless and undamped) of stiffness 250 kN/m, in parallel with a viscous dashpot of rate 20 kN s/m. The 'bouncing' natural frequency of the installation is 2.8 Hz while the two rocking modes are 1.9 Hz and 2.2 Hz.

It is found that excessive high-frequency vibration forces are transmitted from the machine to the floor, particularly above 20 Hz. Some modifications are required to improve the isolation performance, but a constraint is imposed by the pipes and other service connections to the machine which cannot withstand significantly larger displacements than are currently encountered.

It is suggested that a rubber bush be inserted *either* at one end of the dashpot (for example, between the dashpot and the machine structure), *or* between the entire suspension unit and the machine. The same bush would be used in either configuration and it may be modelled as an undamped spring with a stiffness of 700 kN/m.

Show analytically which of the two proposed modifications provides the greatest improvement in high-frequency isolation, and calculate the increased attenuation (in dB) for both cases at 25 Hz and at 50 Hz. Consider motion in the vertical direction only.

Comment on the suitability of the two proposed modifications, and indicate what additional calculations should be made to define completely the dynamic behaviour of the modified installation.

61. A machine having a mass of 1250 kg is isolated from floor vibration by a resilient mount whose stiffness is 0.2×10^6 N/m, and which has negligible damping. The machine generates a strong excitation which can be considered as an externally applied harmonic force at its running speed of 480 revolutions per minute, and the vibration isolation required is specified as a force transmissibility of −35 dB at this frequency.

(i) Show that the single-stage system described above will not provide the necessary attenuation.

It is decided to improve the effectiveness of the installation by introducing a second mass–spring stage between the resilient mount and the floor. The maximum deadweight that can be supported by the floor is 2500 kg, and so the second-stage mass is taken as 1250 kg.

(ii) Calculate the stiffness of the second-stage spring in order to attain the required force transmissibility.

(iii) Determine the frequency at which this two-stage system has the same transmissibility as the simpler single-stage one, and sketch the transmissibility curve for each case, indicating the frequency above which the isolation system gives a definite attenuation.

62. (a) The traditional 'half-power points' formula for estimating damping, that is, loss factor $= \Delta f/f_0$ (where Δf is the frequency bandwidth at the half-power points and f_0 is the frequency of maximum response), is an approximation that becomes unreliable when applied to modes with relatively high damping.

Sketch a graph indicating the error incurred in using this formula instead of the exact one, as a function of damping loss factor in the range 0.1 to 1.0.

(b) The measured receptance data given in the table were taken in the frequency region near a mode of vibration of interest on a scale model of a chemical reactor.

Frequency (Hz)	Receptance	
	Modulus ($\times 10^{-6}$ m/N)	Phase (degrees)
380.0	41.6	31
390.0	49.9	25
400.0	66.5	25
410.0	86.1	41
420.0	70.7	67
430.0	64.0	65
440.0	67.0	60

Obtain estimates for the damping in this mode of vibration, using (a) a modulus–frequency plot and (b) a polar (or Nyquist) plot of the receptance data. Present your answers in terms of Q factors. State which of the two estimates obtained you consider to be the more reliable, and justify your choice.

63. The results given opposite are of an incomplete resonance test on a structure. The response at different frequencies was measured at the point of application of a sinusoidal driving force and is given as the receptance, being the ratio of the amplitude of vibration to the maximum value of the force. The phase angle between the amplitude and force was also measured.

Estimate the effective mass, dynamic stiffness and loss factor, assuming material type damping.

Frequency (Hz)	Receptance ($\times 10^{-6}$ m/N)	Phase angle (degrees)
55	5	0
70	10	3
82	18	24
88	25	40
94	30	55
100	32	85
109	10	135
115	9	180
130	7	180

64. A test is conducted in order to measure the dynamic properties of an antivibration mount. A mass of 900 kg is supported on the mount to form a single degree of freedom system, and measurements are made of the receptance of this system in the region of its major resonance.

In addition to the hysteretic damping provided by the mount (which is to be measured), some additional damping is introduced by friction in the apparatus, and so tests are made at two different amplitudes of vibration (x_0) in order to determine the magnitude of each of the two sources of damping.

It may be assumed that the loss factor of the mount is a constant, valid for all vibration amplitudes, but the dynamic stiffness is not a constant and so the two tests have slightly different natural frequencies.

Details of some receptance measurements are given overleaf. Assuming that the additional damping has the characteristic of Coulomb friction damping, estimate the hysteretic damping loss factor of the mount.

Frequency (Hz)	Receptance measurements			
	Test (a) $x_0 = 0.1$ mm		Test (b) $x_0 = 0.02$ mm	
	$Re(\alpha)$ $(\times 10^{-7}$ m/N$)$	$Im(\alpha)$	$Re(\alpha)$ $(\times 10^{-7}$ m/N$)$	$Im(\alpha)$
13.25	7.6	−6.9	5.1	−4.9
13.50	7.6	−8.6	5.3	−5.6
13.75	6.6	−10.7	5.3	−6.4
14.00	4.4	−12.4	5.2	−7.4
14.25	1.6	−12.8	4.6	−8.5
14.50	−1.0	−11.9	3.7	−9.6
14.75	−2.5	−10.1	2.2	−10.3
15.00	−3.1	−8.4	0.58	−10.2
15.25	−3.2	−7.0	−0.96	−9.6
15.50	−3.0	−6.0	−1.9	−8.5

65. A resonance test on a flexible structure at a constant energy level has revealed a prominent mode at 120 Hz with a half-power frequency bandwidth of 4.8 Hz and a peak acceleration of 480 m/s^2. The effective mass of the structure for this mode has been estimated at 20 kg.

   It is proposed to introduce Coulomb type friction at the point of measurement of the response so as to reduce the motion by 1/5 for the same energy input as before. Estimate the friction force required, assuming this to be independent of amplitude and frequency of vibration.

66. (i) Derive a relationship between the logarithmic decrement of a system with velocity type damping performing free vibrations, and the loss factor for structural damping. Define clearly any assumptions made.

   (ii) A concrete floor slab is to be supported on four columns, spaced in a square grid of sides 7 m. The detail of one column is shown opposite. The slab is to be isolated from vibrations being transmitted up the column by rubber pads, installed as shown. The slab thickness is 150 mm and the density of concrete is 2250 kg/m^3. The first resonance frequency of the slab is estimated to be 20 Hz and the logarithmic decrement for concrete is about 0.2. Measurements of the vibration in a column have shown a strong peak at 20 Hz. The rubber pads have a loss factor of 0.1.

   (a) Estimate the resonance frequency the pad system should have to provide a reduction of 4/5 in the vibration being transmitted at 20 Hz to the centre of the floor slab. Comment upon the result.

   (b) Estimate the additional attenuation in dB the isolation will provide at a frequency of 160 Hz.

   Some of the information given may be superfluous.

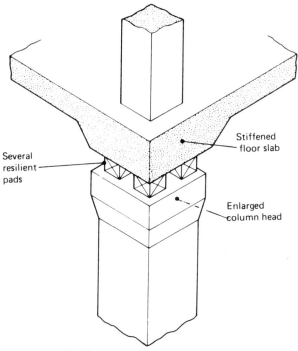

Resilient mounting at elevated level

67. (i) Often it is required to introduce into a structure some additional form of damping
     in order to keep resonance vibrations down to an acceptable level. One method is
     to use a damped dynamic absorber where a (relatively) small mass is suspended
     from the primary mass (of the vibrating structure) by a spring and a dashpot. If
     the absorber spring–mass system is tuned to the natural frequency of the effective
     mass–spring model of the structure, then the absorber dashpot may introduce
     some damping to the structural resonance.
         Without performing analysis, but using physical reasoning only, sketch a
     family of curves for the point receptance on the primary mass for a range of
     different magnitudes for the absorber dashpot between 0 and ∞, and hence show
     that there will be an optimum value for that dashpot rate.
  (ii) In one application of this type of damper-absorber, it is required to increase the
     damping in a new suspension bridge.
         In moderate to high winds the airflow over the bridge generates an effectively
     steady-state excitation force at the bridge's fundamental natural frequency. The
     airflow also provides some damping. The amplitude of steady vibration under
     this excitation is found to be 20 mm and this is twice the maximum amplitude
     considered to be 'safe'. Accordingly, it is proposed to introduce extra damping to
     reduce the resonance amplitude to 10 mm.
         Tests on the bridge show that it possesses some structural damping and this is
     estimated from measurement of free decay curves. The amplitude of vibration is

found to halve after 40 cycles. A more significant source of damping is the airflow over the bridge and this is most readily described in terms of energy dissipation, which is estimated to be βx_0^3 N m per cycle where $\beta = 2.5 \times 10^7$ N/m^2 and x_0 = vibration amplitude. The effective mass and stiffness of the bridge (for its fundamental mode) are 500 000 kg and 5×10^6 N/m, respectively. Determine the equivalent viscous dashpot rate which must be added in order to reduce the resonance vibration amplitude to 10 mm. Assume the excitation force remains the same.

(iii) If, due to miscalculation, the actual dashpot used has a rate of only 70% of that specified, what then will be the vibration amplitude?

68. A partition is made from several layers of metal and a plastic material. Experiments with the metal layers alone have shown that the energy dissipated per cycle of vibration at the lowest natural frequency is $3 \times 10^4 x_0^2$ joule/cycle, where x_0 is the amplitude in metres at the centre of the partition. The stiffness of the plastic layers themselves when measured at the centre is 4×10^5 N/m with a loss factor of 0.3. The acoustic energy loss from one face alone of the partition when vibrating at the lowest natural frequency of 70 Hz is $1.5\dot{x}^2$ joules/cycle, where $\dot{x}$ is the maximum velocity in m/s at the centre.

Calculate the amplitude of vibration at the centre of the partition when one face receives an acoustic energy input of 50 watts at 70 Hz. Explain carefully any assumptions that have to be made.

69. A sketch is given below of the essential parts of the front suspension of a motor car, showing the unsprung mass consisting of the tyre, the wheel, and the stub axle, connected at point A by a rubber bush to a hydraulic shock absorber and the main coil spring. The other end of the shock absorber is connected at point B by another rubber bush to a subframe of the car body. A set of wishbone link arms with rubber bushes at each end serve to stabilize the unit.

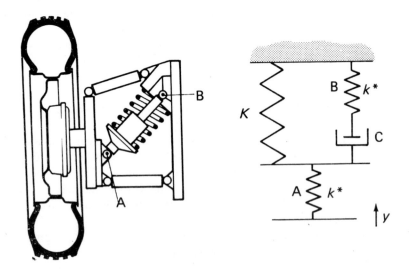

(a) Devise a representative model for this suspension system comprising lumped masses, springs and dampers. Indicate how the equations of motion can be obtained, but do *not* solve these. Define the symbols introduced carefully.

(b) A massless model of the main spring, the shock absorber and the bushes at points A and B is shown, assuming that the car body represents an infinite impedance. The rubber bushes A and B are identical and have a complex stiffness

$$k^* = k(1 + j\eta),$$

where the elastic stiffness $k = 600$ kN/m and the loss factor $\eta = 0.25$.

The main spring stiffness $K = 25$ kN/m, and the shock absorber behaves as a viscous damper with a coefficient $c = 3$ kN s/m.

Estimate the percentage contribution by the two bushes to the total energy being dissipated per cycle, for an input motion $y = y_0 \sin vt$ with $y_0 = 25$ mm and $v = 30$ rad/s, assuming that the maximum possible displacement of 5 mm across each bush is being taken up.

70. A machine produces a vertical harmonic force and is to be isolated from the foundations by a suspension system consisting of metal springs in series with blocks of a viscoelastic material.

(i) Show analytically whether it will be better to place the blocks of viscoelastic material above or below the springs from the point of view of:

(a) the force transmitted to the foundation,

(b) damping out high-frequency resonances in the metal spring for the type of installation where the attachment points to the machine are slender metal brackets.

(ii) Compare the system described above with one in which the blocks of viscoelastic material are placed in parallel with the springs.

Define carefully all symbols introduced.

71. Consider the simple joint shown, in which metal to metal contact occurs.

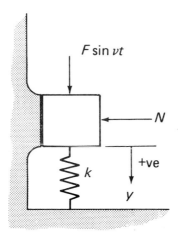

A harmonic exciting force $F \sin vt$ is applied to one joint member which has a mass m, and is supported by an element of stiffness k. The other member is rigidly fixed, so that it is infinitely stiff in the direction of this exciting force. A constant force N is applied normal to the joint interfaces by a clamping arrangement not shown. It is to be assumed that the coefficient of friction μ existing at the joint interfaces is constant.

Assuming the motion y to be sinusoidal, show that

$$y = \frac{F \sin vt - \mu N}{k\left(1 - \left(\dfrac{v}{\omega}\right)^2\right)},$$

and hence obtain an expression for the energy dissipated per cycle by slipping. Show that the maximum energy is dissipated when $\mu N/F = 0.5$, and that y then has an amplitude 50% of the amplitude when $N = 0$. Furthermore, by drawing force–slip hysteresis loops and plotting energy dissipation as a function of $\mu N/F$, show that at least 50% of the maximum energy dissipation can be achieved by maintaining $\mu N/F$ between 0.15 and 0.85.

Comment on the practical significance of this.

72. A beam on elastic supports with dry friction damped joints is modelled by the system shown.

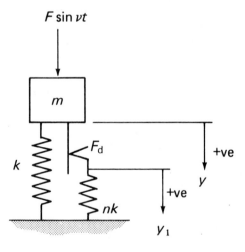

By considering equivalent viscous damping for the friction damper, show that

$$|\delta|^2 = Y^2 - (4F_d/\pi nk)^2,$$

where $\delta = y - y_1$, Y is the amplitude of the body motion and F_d is the tangential friction force in the damper. Hence deduce that

$$Y = \left[\frac{\left(\frac{F}{k}\right)^2 + \left(\frac{4F_d}{\pi nk}\right)^2 \left\{\left[1 - \left(\frac{v}{\omega}\right)^2\right]^2 - \left[1 + n - \left(\frac{v}{\omega}\right)^2\right]^2\right\}}{\left[1 - \left(\frac{v}{\omega}\right)^2\right]^2} \right]^{1/2}$$

Consider the response when $F_d = 0$ and $F_d = \infty$, and show that the amplitude of the body for all values of F_d is $2F/nK$ when $v/\omega = \sqrt{(1 + (n/2))}$, and assess the significance of this.

Hint: Write equations of motion for a system with equivalent viscous damping $c = 4F_d/\pi v$, and put $y_1 = Y_1 e^{jvt}$, etc. From the equations of motion,

$$Y = \frac{F}{k} \left[\frac{1 + (cv/nk)^2}{\left[1 - \left(\frac{v}{\omega}\right)^2\right]^2 + \left(\frac{cv}{nk}\right)^2 \left[1 + n - \left(\frac{v}{\omega}\right)^2\right]^2} \right]^{1/2}.$$

Substituting for c and $|\delta|$ gives the required expression for Y. Note that as $F_d \to \infty$, $|\delta| \to 0$.

73. Part of a structure is modelled by a cantilever with a friction joint at the free end, as shown. The cantilever has a harmonic exciting force $F \sin vt$ applied at a distance a from the root. The tangential friction force generated in the joint by the clamping force N can be represented by a series of linear periodic functions, $F_d(t)$.

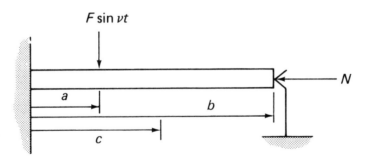

Show that $y_c(t) = \alpha_{ca}F \sin vt + \alpha_{cb}F_d(t)$, where c is an arbitrary position along the cantilever and α is a receptance.

By assuming that the friction force is harmonic and always opposes the exciting force, find the energy dissipated per cycle, and hence show that $F_d = 2\mu N$. Is this assumption reasonable for all modes of vibration?

Thus, this linearization of the damping replaces the actual friction force during slipping, μN, by a sinusoidally varying force of amplitude $2\mu N$. Compare this representation with a Fourier series for the friction damping force.

74. A fabricated steel mast is observed to oscillate violently under certain wind conditions. In order to increase the damping some relative motion is to be allowed in a number of the bolted joints by inserting spring washers under the nuts and by opening

the holes to give a definite clearance. Rubber blocks are to be provided to keep the joint central.

In 15 joints metal to metal sliding friction is to be introduced with a coefficient of friction of 0.2 for a clamping force of 2×10^4 N. The clearance in each bolt hole is 2.5 mm on diameter. To keep the joint nominally at its central position two rubber blocks are fitted as shown below. The blocks are pressed in position to provide a centring force in excess of the static friction force. Each block is square in cross section, 60 mm by 60 mm, and nominally 18 mm thick. The rubber has a loss factor of 0.12.

The maximum energy input per cycle of oscillation of the mast by the wind is estimated as 1500 joules.

(i) Calculate the modulus of elasticity for the rubber material so that the full clearance in the bolt holes of all the joints is just taken up during an oscillation under the maximum wind excitation, neglecting any structural damping in the mast itself.

(ii) Estimate the Q factor for the mast with the damping in the joints, given that the stored energy in the structure is 2500 joules for a deflection which takes up the total clearance in each joint.

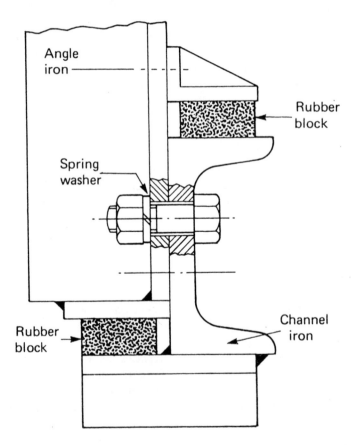

75. The receptance at a point in a structure is measured over a frequency range, and it is found that a resonance occurs in the excitation range. It is therefore decided to add an undamped vibration absorber to the structure.

    Sketch a typical receptance–frequency plot for the structure, and by adding the receptance plot of an undamped vibration absorber, predict the new natural frequencies. Show the effect of changes in the absorber mass and stiffness on the natural frequencies, by drawing new receptance–frequency curves for the absorber.

76. Briefly derive the equations that describe the operation of an undamped dynamic vibration absorber.

    A milling machine of mass 2700 kg demonstrates a large resonant vibration in the vertical direction at a cutter speed of 300 rev/min when fitted with a cutter having 20 teeth. To overcome this effect it is proposed to add an undamped vibration absorber.

    Calculate the minimum absorber mass and the relevant spring stiffness required if the resonance frequency is to lie outside the range corresponding to a cutter speed of 250 to 350 rev/min.

77. In a pumping station, a section of pipe resonated at a pump speed of 120 rev/min. To eliminate this vibration, it was proposed to clamp a spring–mass system to the pipe to act as an absorber. In the first test, an absorber mass of 2 kg tuned to 120 cycle/min resulted in the system having a natural frequency of 96 cycle/min.

    If the absorber system is to be designed so that the natural frequencies lie outside the range 85–160 cycle/min, what are the limiting values of the absorber mass and spring stiffness?

78. A certain machine of mass 300 kg produces a harmonic disturbing force $F \cos 15t$. Because the frequency of this force coincides with the natural frequency of the machine on its spring mounting an undamped vibration absorber is to be fitted.

    If no resonance is to be within 10% of the exciting frequency, find the minimum mass and corresponding stiffness of a suitable absorber. Derive your analysis from the equations of motion, treating the problem as one-dimensional.

79. A machine tool of mass 3000 kg has a large resonance vibration in the vertical direction at 120 Hz. To control this resonance, an undamped vibration absorber of mass 600 kg is fitted, tuned to 120 Hz.

    Find the frequency range in which the amplitude of the machine vibration is less with the absorber fitted than without.

80. The figure overleaf shows a body of mass m_1 which is supported by a spring of stiffness k_1 and which is excited by a harmonic force $P \sin vt$. An undamped dynamic vibration absorber consisting of a mass m_2 and a spring of stiffness k_2 is attached to the body as shown.

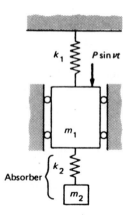

Derive an expression for the amplitude of the vibrations of the body.

The body shows a violent resonance at 152 Hz. As a trial remedy a vibration absorber is attached which results in a resonance frequency at 140 Hz. How many such absorbers are required if no resonance is to occur between 120 and 180 Hz?

81. (i) Show that the frequency at which an undamped vibration absorber is most effective (ω_a) is given by the expression

$$\omega_a = \frac{k_a}{m_a}$$

(where m_a and k_a are the mass and stiffness of the added absorber system) and is therefore independent of the properties of the system to which the absorber has been added. Also, derive an expression for the steady-state amplitude of the absorber mass when the system is being driven at its natural frequency ω_a.

(ii) In order to suppress vibration, a vibration absorber system is to be attached to a machine tool which operates over a range of speeds. The design of the absorber is chosen to be a light beam, which is rigidly fixed at one end to the machine tool, and a mass, which may be clamped at various positions along the length of the beam so as to tune the absorber to a required frequency.

Given that the beam is made of aluminium which has a Young's Modulus of 70 GN/m² and is of square section, 60 mm × 60 mm, and the absorber mass is 25 kg, calculate the minimum length of the beam required for the absorber to function over the frequency range 40–50 Hz. Ignore the mass of the beam itself.

Also, calculate how far from the fixed end of the beam the mass would have to be clamped in order to tune the absorber to the maximum frequency of its range of operation.

7

Answers and solutions to selected problems

1.

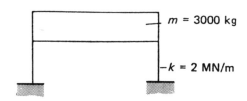

m = 3000 kg

$-k$ = 2 MN/m

$$f = \frac{1}{2\pi}\sqrt{\left(\frac{k}{m}\right)} = \frac{1}{2\pi}\sqrt{\left(\frac{2 \times 2 \times 10^6}{3000}\right)} = \underline{5.8 \text{ Hz.}}$$

2.   FBDs are

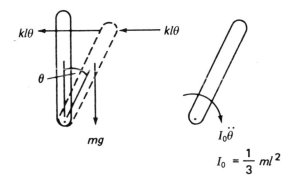

$kl\theta$ $kl\theta$

θ

mg

$I_0\ddot{\theta}$

$I_0 = \frac{1}{3}ml^2$

Moments about pivot gives

$$\tfrac{1}{3}ml^2\ddot\theta = mg\,\frac{l}{2}\,\theta - 2kl^2\theta,$$

so

$$\ddot\theta + \theta\left[\frac{2kl^2 - mgl/2}{\tfrac{1}{3}ml^2}\right] = 0$$

and

$$f = \frac{1}{2\pi}\sqrt{\left(\frac{12kl - 3mg}{2ml}\right)}\ \text{Hz.}$$

3. <u>3.6 Hz.</u>

4. $$I = \frac{\pi}{64}(80^4 - 70^4) = 832 \times 10^3\ \text{mm}^4,$$

and

$$\text{mass/length} = \frac{\pi}{4}(0.08^2 - 0.07^2)\,7750 + \frac{\pi}{4}(0.07)^2\,930$$

$$= 12.71\ \text{kg/m.}$$

Now

$$\omega^2 = g\,\frac{\displaystyle\int_0^L x^2(L-x)^2\,dx}{(mg/24EI)\displaystyle\int_0^L x^4(L-x)^4\,dx},$$

$$= \frac{24EI}{M}\,\frac{L^5/30}{L^9/630} = \frac{24EI}{m}\,\frac{21}{L^4},$$

$$= \frac{24 \times 200 \times 10^9 \times 832 \times 10^3 \times 21}{12.71 \times 10^{12} \times 4^4}\,\text{s}^{-2}$$

Hence $\omega = 161\ \text{rad/s}$ and $\underline{f = 25.6\ \text{Hz.}}$

5.

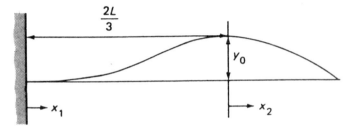

Assume

$$\text{for } 0 < x_1 < \frac{2L}{3}, \quad y = \frac{y_0}{2}\left(1 - \cos\frac{3\pi}{2L}x_1\right),$$

and

$$\text{for } 0 < x_2 < \frac{L}{3}, \quad y = y_0 \cos\frac{3\pi}{2L}x_2.$$

Now

$$\omega^2 = \frac{\displaystyle\int EI\,(d^2y/dx^2)^2\,dx}{\displaystyle\int y^2\,dm},$$

where

$$\int_0^L \left(\frac{d^2y}{dx^2}\right)^2 dx$$

$$= y_0^2\left[\int_0^{2L/3}\frac{1}{4}\left(\frac{3\pi}{2L}\right)^4\cos^2\frac{3\pi}{2L}x_1\,dx_1 + \int_0^{L/3}\left(\frac{3\pi}{2L}\right)^4\cos^2\frac{3\pi}{2L}x_2\,dx_2\right]$$

$$= y_0^2\left(\frac{3\pi}{2L}\right)^4\frac{L}{4},$$

and

$$\int_0^L y^2\,dm = my_0^2\left[\int_0^{2L/3}\frac{1}{4}\left(1 - 2\cos\frac{3\pi}{2L}x_1 + \cos^2\frac{3\pi}{2L}x_1\right)dx_1\right.$$

$$\left. + \int_0^{L/3}\cos^2\frac{3\pi}{2L}x_2\,dx_2\right]$$

$$= \frac{5mL}{12}\, y_0^2.$$

Substituting numerical quantities gives $\underline{f = 1.65 \text{ Hz}}$.

6.   1.45 Hz.

7.   8.5 Hz.

9.

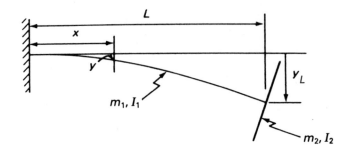

Assume

$$y = y_L\left(1 - \cos\left(\frac{\pi}{2}\frac{x}{L}\right)\right),$$

then

$$\frac{d^2y}{dx^2} = -\left(\frac{\pi}{2L}\right)^2 \cos\left(\frac{\pi}{2}\frac{x}{L}\right).$$

Now

$$\omega^2 = \frac{\displaystyle\int EI\,(d^2y/dx^2)^2\,dx}{\displaystyle\int y^2\,dm},$$

where

$$\int_0^L EI\left(\frac{d^2y}{dx^2}\right)^2 dx = EI_1 \int_0^L \left(\frac{\pi}{2L}\right)^4 \cos^2\left(\frac{\pi}{2}\frac{x}{L}\right) dx$$

$$= EI_1\left(\frac{\pi}{2L}\right)^4 y_L^2 \frac{L}{2}$$

and

$$\int y^2 \, dm = \int_0^L \frac{m_1}{L} y^2 \, dx + m_2 y_L^2 + I_2 \left(\frac{dy_L}{dx}\right)^2$$

$$= \frac{m_1}{L} y_L^2 \int_0^L \left[1 - 2\cos\left(\frac{\pi}{2}\frac{x}{L}\right) + \cos^2\left(\frac{\pi}{2}\frac{x}{L}\right)\right] dx + m_2 y_L^2$$

$$+ I_2 \left(\frac{\pi}{2L}\right)^2 y_L^2$$

$$= \frac{m_1}{L} y_L^2 \left[L - \frac{4L}{\pi} + \frac{L}{2}\right].$$

Substituting numerical values gives

$$E = 207 \times 10^9 \text{ N/m}^2,$$

$$L = 0.45 \text{ m},$$

$$I_1 = \frac{\pi}{64} \times 2.5^4 \times 10^{-8} = 1.916 \times 10^{-8} \text{ m}^4,$$

$$m_1 = 7850 \times \frac{\pi}{4} \times (0.025)^2 \times 0.45 = 1.732 \text{ kg},$$

$$m_2 = 7850 \times \pi \times 0.58(0.02)^2 = 5.71 \text{ kg}$$

and

$$I_2 = \tfrac{1}{2} \times 5.71 \times \left(\frac{0.58}{2}\right)^2 = 0.24 \text{ kg m}^2,$$

so that

$$\omega^2 = \frac{1.325 \times 10^5}{9.026},$$

and $\omega = 121$ rad/s, so that $\underline{f = 19.3 \text{ Hz}}$.

10. 9.8 Hz.

12. 5.5 s; 65 m.

14. $$\Lambda = \frac{1}{10} \ln \frac{100}{1} = 0.46$$

$$\simeq 2\pi\zeta, \quad \text{so} \quad \zeta = 0.0732.$$

$$\omega_v = \omega\sqrt{(1 - \zeta^2)}, \quad \text{so } f_v = f\sqrt{(1 - \zeta^2)},$$

and

$$\tau = \tau_v \sqrt{(1 - \zeta^2)}$$
$$= 1 \sqrt{(1 - 0.0732^2)}$$
$$= \underline{0.997 \text{ s.}}$$

15. In 2 s execute 3 cycles, so

$$\Lambda = \tfrac{1}{3} \ln \frac{1}{0.9} = \underline{0.035.}$$

For small Λ, $\Lambda \simeq 2\pi\zeta$, hence $\underline{\zeta = 0.00557.}$

Also for small ζ, $\omega \simeq \omega_v$

so

$$c_c = 2\sqrt{(km)} = 2m\omega = 2 \times 10^5 \times 3\pi = 1885 \times 10^3 \text{ N m s/rad};$$
$$c = \zeta c_c = 0.00557 \times 1885 \times 10^3 = \underline{10\,500 \text{ N m s/rad.}}$$

16. $\zeta = 0$ so $\dfrac{X_T}{X_0} = \dfrac{1}{1 - (v/\omega)^2} = 0.1.$

Hence

$$\frac{v}{\omega} = 3.32.$$

Since

$$v = 80\pi \text{ rad/s}, \quad \omega = \frac{80\pi}{3.32} = \sqrt{\left(\frac{g}{\delta}\right)},$$

so that

$$\delta = \left(\frac{3.32}{80\pi}\right)^2 \times 9.81 = \underline{1.7 \text{ mm.}}$$

17. $\dfrac{f_1}{f_2} = \dfrac{(1/2\pi)\sqrt{(k/m)}}{(1/2\pi)\sqrt{(k/(m + M))}}.$

Thus

$$\frac{f_1}{f_2} = \sqrt{\left(\frac{m + M}{m}\right)} = \frac{\tau_2}{\tau_1},$$

so

$$\frac{0.62}{0.6} = \sqrt{\left(\frac{m + 4000}{m}\right)},$$

and

$\underline{m = 59\ 000\ \text{kg}}.$

18. $$\frac{X_T}{X_0} = \frac{\sqrt{[1 + (2\zeta\ v/\omega)^2]}}{\sqrt{\{[1 - (v/\omega)^2]^2 + [2\zeta\ v/\omega]^2\}}} = \frac{1}{10}.$$

Substituting $\zeta = 0.2$ gives

$$\left(\frac{v}{\omega}\right)^4 - 17.84 \left(\frac{v}{\omega}\right)^2 - 99 = 0,$$

so that

$$\left(\frac{v}{\omega}\right) = 4.72.$$

Limit at 15 Hz, so $v = 30\pi$ rad/s and

$$\omega = \frac{30\pi}{4.72} = 19.95 = \sqrt{\left(\frac{k_T}{m}\right)}.$$

Hence

$k_T = 15.92$ kN/m

and

$$k = \frac{k_T}{3} = \underline{5.3\ \text{kN/m}}.$$

19. If n units are required, then

$k = 359 \times 10^3 n$ N/m,

$c = 2410n$ N s/m

$\omega^2 = 359 \times 10^3 n/520 = 690.4\ n\ \text{s}^{-2}$

$c_c = 2\sqrt{(km)},$

so

$c_c^2 = 4 \times 359 \times 520 \times 10^3 n = 747 \times 10^6 n.$

$v = 25 \times 2\pi$ rad/s,   so $v^2 = 2.46 \times 10^4/\text{s}^2.$

Hence

$$\frac{F_T}{F_0} = 0.4, \quad \text{and} \quad \frac{v^2}{\omega^2} = 35.6/n.$$

Substitute values in

$$\left(\frac{F_T}{F_0}\right)^2 = \frac{1 + \left(2 \dfrac{c}{c_0} \dfrac{v}{\omega}\right)^2}{\left[\left(1 - \dfrac{v^2}{\omega^2}\right)^2 + \left(2 \dfrac{c}{c_0} \dfrac{v}{\omega}\right)^2\right]}$$

to give

$$n^2 + 6.42n - 114.6 = 0,$$

so

$$n = +8 \text{ or } -14.5$$

Thus 8 units in parallel will give specified attenuation; more units would give less static deflection but more transmitted force.

20.

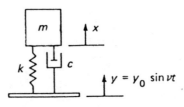

Equation of motion is $m\ddot{x} = k(y - x) + c(\dot{y} - \dot{x})$. If

$$z = y - x, \quad \dot{z} = \dot{y} - \dot{x}, \quad \text{and} \quad \ddot{z} = \ddot{y} - \ddot{x},$$

then

$$m\ddot{z} + c\dot{z} + kz = m\ddot{y} = -mv^2 y_0 \sin vt.$$

Assume

$$z = z_0 \sin(vt + \phi),$$

then

$$-mv^2 z_0 \sin(vt + \phi) + cz_0 v \cos(vt + \phi) + kz_0 \sin(vt + \phi)$$
$$= -mv^2 y_0 \sin vt.$$

If

$$\omega = \sqrt{\left(\frac{k}{m}\right)} \quad \text{and} \quad \zeta = \frac{c}{c_c} = \frac{c\omega}{2k},$$

then

$$z_0 = \frac{(v/\omega)^2 \, y_0}{\sqrt{\left\{\left[1 - \left(\dfrac{v}{\omega}\right)^2\right]^2 + \left[2\zeta\dfrac{v}{\omega}\right]^2\right\}}}.$$

Now if $\omega \gg v$, $(v/\omega)^2 \ll 1$ and

$$z_0 \simeq \left(\frac{v}{\omega}\right)^2 y_0,$$

that is, the acceleration $v^2 y_0$ is measured.

As (v/ω) increases, $[1 - (v/\omega)^2]^2$ decreases, but the damping term in the denominator increases to compensate. If $v/\omega = 0.2$ and no error is required,

$$\frac{1}{\left[1 - \left(\dfrac{v}{\omega}\right)^2\right]^2 + \left[2\zeta\dfrac{v}{\omega}\right]^2} = 1, \quad \text{when} \quad \frac{v}{\omega} = 0.2,$$

so that

$$[1 - 0.04]^2 + 4\zeta^2[0.04] = 1,$$

and hence $\underline{\zeta = 0.7}$.

21. $x = 0.056$ m; $\phi = 3.7°$.

22. $f(t) = -\dfrac{8}{\pi^2}\left[\cos t + \dfrac{1}{9}\cos 3t + \dfrac{1}{25}\cos 5t + \ldots\right]$.

24. $I_0 = m_1 h^2 + \dfrac{m_2 r^2}{4} + m_2 s^2$.

Equation of motion is

$$I_0\ddot{\theta} + cp^2\dot{\theta} + kq^2\theta = 0.$$

When $c = 0$,

$$\omega = \sqrt{\left(\frac{kq^2}{I_0}\right)} \text{ rad/s.}$$

With damping $\omega_v = \omega\sqrt{(1 - \zeta^2)}$, where

$$\zeta = \frac{c}{c_c} = \frac{cp^2}{2\sqrt{(kq^2 I_0)}}.$$

Thus

$$\omega_v = \omega \sqrt{\left[1 - \left(\frac{c^2 p^4}{4kq^2 I_0}\right)\right]} \text{ rad/s.}$$

25. FBDs are

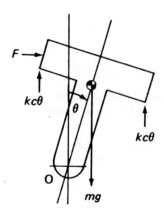 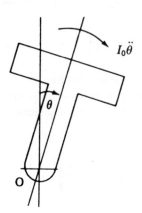

Equation of motion is $Fd + mgh\theta - 2kc^2\theta = I_0\ddot{\theta}$, or

$$\ddot{\theta} + \left(\frac{2kc^2 - mgh}{I_0}\right)\theta = \frac{Fd}{I_0}.$$

C.F. + P.I. give solution as

$$\theta = A \sin \omega t + B \cos \omega t + \frac{Fd}{\omega^2 I_0},$$

where

$$\omega = \sqrt{[(2kc^2 - mgh)/I_0]}.$$

Substitute initial conditions for

$$\theta = \frac{Fd}{\omega^2 I_0}(1 - \cos \omega t).$$

26. Equation of motion is

$$m\ddot{x} + kx = m_r r v^2 \sin vt,$$

so the amplitude of the motion is $m_r r v^2/(k - mv^2)$. Thus

$$k\left(\frac{m_r r v^2}{k - mv^2}\right) = \text{transmitted force.}$$

Substituting values gives

$$k\left(\frac{0.2 \times 0.01 \times 150^2}{k - (2 \times 150^2)}\right) = 100.$$

Hence $k = 81.8 \text{ kN/m}$.

27. $m^2\omega^4 - 3mk\omega^2 + k^2 = 0; \sqrt{\left(\frac{k}{m}\right)}, \sqrt{\left(\frac{3 \pm \sqrt{5}}{2}\right)}.$

28. System is

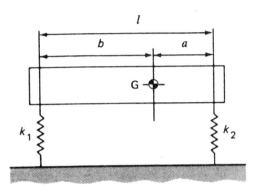

where

$$m = 2000 \text{ kg}; l = 3 \text{ m}; a = 1 \text{ m}; b = 2\text{m}; I_G = 500 \text{ kg m}^2;$$
$$k_2 = 50 \times 10^3 \text{ N/m}; k_1 = 80 \times 10^3 \text{ N/m}.$$

FBDs:

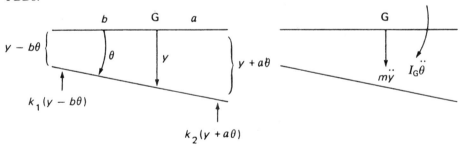

Equations of motion are

$$m\ddot{y} = -k_1(y - b\theta) - k_2(y + a\theta),$$

and

$$I_G\ddot{\theta} = k_1(y - b\theta)b - k_2(y + a\theta)a.$$

Substitute

$$y = Y \sin \omega t \quad \text{and} \quad \theta = \Theta \sin \omega t$$

and rearrange:

$$(k_1 + k_2 - m\omega^2)Y + (k_2a - k_1b)\Theta = 0,$$

and

$$(k_2a - k_1b)Y + (k_1b^2 + k_2a^2 - I_G\omega^2)\Theta = 0.$$

Hence the frequency equation is

$$(k_1 + k_2 - m\omega^2)(k_1b^2 + k_2a^2 - I_G\omega^2) - (k_2a - k_1b)^2 = 0.$$

Substituting numerical values and dividing by 10^3 gives

$$(130 - 2\omega^2)(370 - 0.5\omega^2) - (-110)^2 = 0,$$

or

$$\omega^4 - 805\omega^2 + 36\,000 = 0.$$

Hence

$$\omega^2 = \frac{805 \pm 710}{2} = 758 \text{ or } 47.5,$$

so that

$$\underline{f_2 = 4.38 \text{ Hz and } f_1 = 1.10 \text{ Hz.}}$$

The mode shape is obtained from

$$\frac{Y}{\Theta} = \frac{k_1b - k_2a}{k_1 + k_2 - m\omega^2},$$

so that at f_1, $\omega^2 = 47.5$ and

$$\frac{Y}{\Theta} = \frac{160 \times 10^3 - 50 \times 10^3}{130 \times 10^3 - 2 \times 10^3 \times 47.5} = \underline{3.14},$$

and at f_2, $\omega^2 = 758$ and

$$\frac{Y}{\Theta} = -\frac{110}{1386} = \underline{-0.079}.$$

Speeds are $\underline{V_P = 78 \text{ km/h}}$ and $\underline{V_T = 19.8 \text{ km/h.}}$

29. Equations of motion are, for free vibration,

$$(k_1 - m_1\omega^2)\,X_1 + (-k_1)\,X_2 = 0,$$

and

$$(-k_1)\,X_1 + (k_1 + k_2 - m_2\omega^2)\,X_2 = 0.$$

Hence frequency equation is

$$(k_1 - m_1\omega^2)(k_1 + k_2 - m_2\omega^2) - (-k_1)^2 = 0,$$

or

$$\omega^4 \,(m_1 m_2) - \omega^2 \,(m_1 k_1 + m_1 k_2 + m_2 k_1) + k_1 k_2 = 0.$$

Now

$$m_1 = \tfrac{1}{2} m_2 \quad \text{and} \quad k_1 = \tfrac{1}{2} k_2,$$

Thus

$$2 m_1^2 \omega^4 - 5 m_1 k_1 \omega^2 + 2 k_1^2 = 0,$$

or

$$(2 m_1 \omega^2 - k_1)(m_1 \omega^2 - 2 k_1) = 0;$$

that is,

$$\omega^2 = \frac{k_1}{2 m_1} \quad \text{or} \quad \frac{2 k_1}{m_1},$$

so that

$$f_1 = \frac{1}{2\pi} \sqrt{\left(\frac{k_1}{2 m_1}\right)} \text{ Hz} \quad \text{and} \quad f_2 = \frac{1}{2\pi} \sqrt{\left(\frac{2 k_1}{m_1}\right)} \text{ Hz}.$$

Now

$$\frac{X_1}{X_2} = \frac{k_1}{k_1 - m_1 \omega^2};$$

that is, at frequency f_1,

$$\frac{X_1}{X_2} = + \, 0.5$$

and at frequency f_2,

$$\frac{X_1}{X_2} = - \, 1.0.$$

With harmonic force $F_1 \sin vt$ applied,

$$(k_1 - m_1 \omega^2) X_1 + (-k_1) X_2 = F_1,$$

and

$$(-k_1) X_1 + (k_1 + k_2 - m_2 \omega^2) X_2 = 0.$$

Hence

$$X_1 = \frac{(k_1 + k_2 - m_2 \omega^2)}{\Delta} F_1$$

and

$$X_2 = \frac{k_1}{\Delta} F_1$$

where

$$\Delta = m_1 m_2 \omega^4 - \omega^2(m_1 k_1 + m_1 k_2 + m_2 k_1) + k_1 k_2.$$

30. FBD:

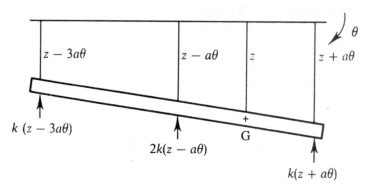

Equations of motion are

$$-k(z - 3a\theta) - 2k(z - a\theta) - k(z + a\theta) = m\ddot{z},$$

and

$$k(z - 3a\theta)3a + 2k(z - a\theta)a - k(z + a\theta)a = I_G\ddot{\theta}.$$

Substitute

$$I = 2ma^2, z = A \sin \omega t, \quad \text{and} \quad \theta = B \sin \omega t,$$

to give

$$(m\omega^2 - 4k)A + 4kaB = 0,$$

and

$$4kaA + (2ma^2\omega^2 - 12ka^2)B = 0,$$

so that the frequency equation is

$$(m\omega^2 - 4k)(2ma^2\omega^2 - 12ka^2) - 16k^2a^2 = 0.$$

Multiply out and factorize to give

$$f_1 = \frac{1}{2\pi} \sqrt{\left(\frac{2k}{m}\right)} \text{ Hz } \quad \text{and} \quad f_2 = \frac{1}{2\pi} \sqrt{\left(\frac{8k}{m}\right)} \text{ Hz.}$$

31. FBDs are as below:

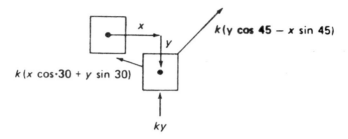

that is,

$$\frac{kx\sqrt{3}}{4} + \frac{ky}{4} \qquad \frac{ky}{2} - \frac{kx}{2}$$

$$\frac{kx3}{4} + \frac{ky\sqrt{3}}{4} \qquad \qquad \frac{ky}{2} - \frac{kx}{2} \qquad \qquad m\ddot{x}$$

$$ky \qquad\qquad\qquad m\ddot{y}$$

Equations of motion are

$$(\Sigma F_x) \quad m\ddot{x} = \frac{ky}{2} - \frac{kx}{2} - kx\frac{3}{4} - ky\frac{\sqrt{3}}{4}$$

and

$$(\Sigma F_y) \quad m\ddot{y} = -ky - \frac{ky}{2} + \frac{kx}{2} - kx\frac{\sqrt{3}}{4} - \frac{ky}{4}.$$

Assuming a solution of the form $x = X \sin \omega t$, $y = Y \sin \omega t$, these are

$$-m\omega^2 X + \frac{5k}{4} X + \left(\frac{\sqrt{3}-2}{4}\right) kY = 0,$$

and

$$\left(\frac{\sqrt{3}-2}{4}\right) kX - m\omega^2 Y + \frac{7}{4} kY = 0.$$

Hence

$$\left(\frac{5}{4}k - m\omega^2\right)\left(\frac{7}{4}k - m\omega^2\right) - \left(\left(\frac{\sqrt{3}-2}{4}\right)k\right)^2 = 0$$

and

$$m^2\omega^4 - 3mk\omega^2 + \frac{k^2}{16}(28 + 4\sqrt{3}) = 0.$$

Hence

$$\frac{\omega^4}{\Omega^4} - 3\frac{\omega^2}{\Omega^2} + 2.183 = 0,$$

where

$$\Omega = \sqrt{\left(\frac{k}{m}\right)}$$

and

$$\frac{\omega^2}{\Omega^2} = 1.326 \text{ or } 1.114.$$

Thus frequencies of vibration are

$$\frac{1.11}{2\pi}\sqrt{\left(\frac{k}{m}\right)} \text{ Hz} \quad \text{and} \quad \frac{1.33}{2\pi}\sqrt{\left(\frac{k}{m}\right)} \text{ Hz.}$$

33. FBDs are

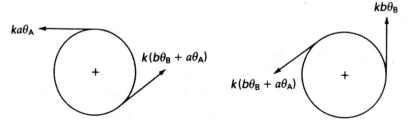

The equations of motion are

$$-ka^2\theta_A - kab\theta_B - ka^2\theta_A = I_A\ddot{\theta}_A$$

and

$$-kab\theta_A - kb^2\theta_B - kb^2\theta_B = I_B\ddot{\theta}_B.$$

Substitute $\theta_A = A \sin \omega t$ and $\theta_B = B \sin \omega t$ to give

$$(-I_A\omega^2 + 2ka^2)A + kab\, B = 0,$$

and

$$kab\, A + (-I_A\omega^2 + 2kb^2)B = 0.$$

So the frequency equation is

$$I_A I_B \omega^4 - 2k(I_B a^2 + I_A b^2)\omega^2 + 3k^2 a^2 b^2 = 0.$$

Substitute numerical values to give

$$\omega = \underline{19.9 \text{ rad/s}} \quad \text{or} \quad \underline{35.7 \text{ rad/s}}.$$

At 19.9 rad/s, $A/B = -1.65$
and at 35.7 rad/s $A/B = +3.68$.

34. $T = \frac{1}{2}m_1 L^2 \dot{\theta}_1^2 + \frac{1}{2}m_2 L^2 \dot{\theta}_2^2 + \frac{1}{2}m_3 L^2 \dot{\theta}_3^2,$

$$V = m_1 gL(1 - \cos \theta_1) + m_2 gL(1 - \cos \theta_2) + m_3 gL(1 - \cos \theta_3)$$
$$+ \tfrac{1}{2}k_1(a \sin \theta_2 - a \sin \theta_1)^2 + \tfrac{1}{2}k_2(a \sin \theta_3 - a \sin \theta_2)^2.$$

For small oscillations,

$$1 - \cos \theta \simeq \frac{\theta^2}{2} \quad \text{and} \quad \sin \theta \simeq \theta.$$

Apply Lagrange equation with $q_i = \theta_1, \theta_2, \theta_3$ in turn to obtain the equations of motion.

35. FBDs:

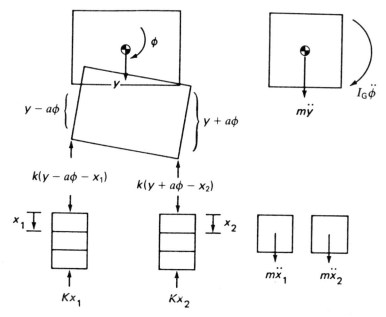

Equations of motion are:

$$m\ddot{y} = -k(y - a\phi - x_1) - k(y + a\phi - x_2),$$
$$I_G \ddot{\phi} = k(y - a\phi - x_1)a - k(y + a\phi - x_2)a,$$
$$m\ddot{x}_1 = k(y - a\phi - x_1) - Kx_1$$

and

$$m\ddot{x}_2 = k(y + a\phi - x_2) - Kx_2.$$

Assume for wheel hop that body does not move; then

$$y = \phi = 0 \quad \text{and} \quad f = \frac{1}{2\pi}\sqrt{\left(\frac{K + k}{m}\right)},$$

$$= \frac{1}{2\pi}\sqrt{\left(\frac{90 \times 10^3}{22.5}\right)} = \underline{10.1 \text{ Hz.}}$$

37. $$f_1 = \frac{1}{2\pi}\sqrt{\left(\frac{3k}{m} - \frac{\sqrt{3}k}{m}\right)} \text{ Hz};$$

$$f_2 = \frac{1}{2\pi}\sqrt{\left(\frac{3k}{m}\right)} \text{ Hz};$$

$$f_3 = \frac{1}{2\pi}\sqrt{\left(\frac{3k}{m} + \frac{\sqrt{3}k}{m}\right)} \text{ Hz.}$$

40. Consider half of aircraft:

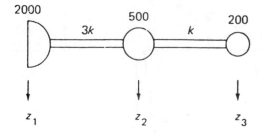

Equations of motion are

$$3k(z_2 - z_1) = 2000 \ddot{z},$$

$$3k(z_1 - z_2) + k(z_3 - z_2) = 500 \ddot{z}_2,$$

and

$$k(z_2 - z_3) = 200 \ddot{z}_3.$$

Substitute

$$z_1 = A_1 \sin \omega t, \quad z_2 = A_2 \sin \omega t \quad \text{and} \quad z_3 = A_3 \sin \omega t,$$

and eliminate A_1, A_2, A_3 to give frequency equation as

$$2 \times 10^4 \omega^4 - 290 \omega^2 k + 0.81 k^2 = 0.$$

Hence

$$\omega^2 = 379 \quad \text{or} \quad 1074,$$

and

$$f_1 = 3.10 \text{ Hz} \quad \text{and} \quad f_2 = 5.22 \text{ Hz}.$$

41. Model system as follows:

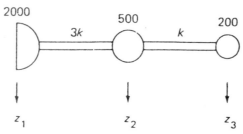

2000

500

200

3k

k

z_1

z_2

z_3

Equations of motion are

$$I_1 \ddot{\theta}_1 = -k_1(\theta_1 - \theta_2),$$

$$I_2 \ddot{\theta}_2 = k_1(\theta_1 - \theta_2) - k_2(\theta_2 - \theta_3)$$

and

$$I_3 \ddot{\theta}_3 = k_2(\theta_2 - \theta_3).$$

Substituting $\theta_i = \Theta_i \sin \omega t$ gives

$$
\begin{aligned}
\Theta_1[k_1 - I_2\omega^2] + \Theta_2[-k_1] &= 0, \\
\Theta_1[-k_1] \quad + \Theta_2[k_1 + k_2 - I_2\omega^2] + \Theta_3[-k_2] &= 0
\end{aligned}
$$

and

$$\Theta_2[-k_2] \quad\quad\quad + \Theta_3[k_2 - I_3\omega^2] = 0.$$

The frequency equation is therefore

$$(k_1 - I_1\omega^2)[(k_1 + k_2 - I_2\omega^2)(k_2 - I_3\omega^2) - k_2^2] + k_1[(k_2 - I_3\omega^2)(-k_1)] = 0;$$

that is,

$$\omega^2[I_1 I_2 I_3 \omega^4 - \omega^2(k_1 I_2 I_3 + k_1 I_1 I_3 + k_2 I_1 I_3 + k_2 I_1 I_2)$$
$$+ k_1 k_2 (I_1 + I_2 + I_3)] = 0,$$

so that either $\omega = 0$ (rigid body rotation) or [...] = 0.
Substituting numerical values gives

$$0.48 \omega^4 - 1100 \times 10^3 \omega^2 + 149 \times 10^9 = 0,$$

so that

$$\omega_1 = 380 \text{ rad/s} \quad \text{and} \quad \omega_2 = 1460 \text{ rad/s};$$

that is,

$$f_1 = 60.5 \text{ Hz} \quad \text{and} \quad f_2 = 232 \text{ Hz}.$$

At $f_1 = 60.5$ Hz,

$$\frac{\theta_1}{\theta_2} = \frac{k_1}{k_1 - I_1\omega^2} = + 1.4$$

and

$$\frac{\theta_3}{\theta_2} = \frac{k_2}{k_2 - I_3\omega^2} = - 0.697.$$

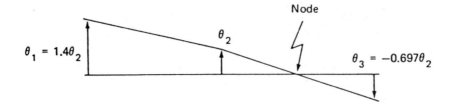

and at $f_2 = 232$ Hz,

$$\frac{\theta_1}{\theta_2} = - 0.304$$

and

$$\frac{\theta_3}{\theta_2} = - 0.028.$$

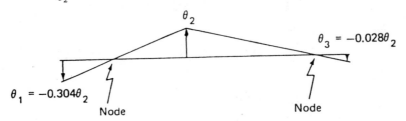

42. $T = \frac{1}{2}m_1\dot{x}_1^2 + \frac{1}{2}m_2\dot{x}_2^2 + \frac{1}{2}m\left(\frac{\dot{x}_1}{2}\right)^2 2 + \frac{1}{2}m\left(\frac{\dot{x}_2}{2}\right)^2 + \frac{1}{2}J\left(\frac{\dot{x}_1}{d}\right)^2 2 + \frac{1}{2}J\left(\frac{\dot{x}_2}{d}\right)^2$

and

$$V = \frac{1}{2}kx_1^2 + \frac{1}{2} \cdot 4k\left(\frac{x_1}{2}\right)^2 + \frac{1}{2}k(x_1 - x_2)^2.$$

Apply the Lagrange equation.

$$\frac{d}{dt}\left(\frac{\partial T}{\partial \dot{x}_1}\right) + \frac{\partial V}{\partial x_1} = 0,$$

$$\frac{d}{dt}\left(\frac{\partial T}{\partial \dot{x}_1}\right) = m_1\ddot{x}_1 + \frac{m}{2}\ddot{x}_1 + \frac{2J}{d^2}\ddot{x}_1,$$

and

$$\frac{\partial V}{\partial x_1} = kx_1 + kx_1 + \tfrac{1}{2}k(2x_1 - x_2) = 3kx_1 - kx_2.$$

Hence equation of motion is

$$\ddot{x}_1\left(m_1 + \frac{m}{2} + \frac{2J}{d^2}\right) + 3kx_1 - kx_2 = 0.$$

Similarly, other equation of motion is

$$\ddot{x}_2\left(m_2 + \frac{m}{4} + \frac{2J}{d^2}\right) + k(x_2 - x_1) = 0.$$

If $m_1 = 4m$, $m_2 = 2m$ and $J = md^2/8$, equations become

$$\ddot{x}_1\left(4m + \frac{m}{2} + \frac{m}{4}\right) + 3kx_1 - kx_2 = 0$$

and

$$\ddot{x}_2\left(2m + \frac{m}{4} + \frac{m}{8}\right) + kx_2 - kx_1 = 0.$$

Assume $x_i = X_i \sin \omega t$, so that

$$X_1\left[3k - \frac{19}{4}m\omega^2\right] + X_2[-k] = 0$$

and

$$X_1[-k] + X_2\left[k - \frac{19}{8}m\omega^2\right] = 0.$$

The frequency equation is therefore

$$\left(3k - \frac{19}{4}m\omega^2\right)\left(k - \frac{19}{8}m\omega^2\right) - (-k)^2 = 0,$$

which is

$$361\left(\frac{\omega}{\Omega}\right)^4 - 380\left(\frac{\omega}{\Omega}\right)^2 + 64 = 0,$$

where

$$\Omega = \sqrt{\left(\frac{k}{m}\right)}.$$

Hence

$$\left(\frac{\omega}{\Omega}\right)^2 = \frac{380 \pm \sqrt{(380^2 - 4 \times 361 \times 64)}}{361 \times 2},$$

and

$$\omega = 0.918\sqrt{\left(\frac{k}{m}\right)} \quad \text{or} \quad 0.459\sqrt{\left(\frac{k}{m}\right)} \text{ rad/s.}$$

For the mode shape,

$$\frac{X_1}{X_2} = \frac{k}{3k - \frac{19}{4}m\omega^2}.$$

When

$$\omega = 0.459\sqrt{\left(\frac{k}{m}\right)} \text{ rad/s,} \quad \frac{X_1}{X_2} = \frac{k}{3k - \frac{19}{4} \times 0.21k} = \underline{+0.5},$$

and when

$$\omega = 0.918\sqrt{\left(\frac{k}{m}\right)} \text{ rad/s,} \quad \frac{X_1}{X_2} = \frac{k}{3k - \frac{19}{4} \times 0.843k} = \underline{-1.0}.$$

43. Assume $x_1 > x_2 > x_3 > x_4$. FBDs are then as follows:

The equations of motion are therefore

$$k_1 x_1 + k_2(x_1 - x_2) + k_3(x_1 - x_3) = -m_1 \ddot{x}_1,$$

$$-k_2(x_1 - x_2) + k_4(x_2 - x_3) = -m_2 \ddot{x}_2,$$

$$-k_3(x_1 - x_3) - k_4(x_2 - x_3) + k_5(x_3 - x_4) = -m_3 \ddot{x}_3,$$

and

$$-k_5(x_3 - x_4) = -m_4 \ddot{x}_4.$$

Substitute $x_i = X_i \sin \omega t$:

$$k_1 X_1 + k_2(X_1 - X_2) + k_3(X_1 - X_3) = m_1 \omega^2 X_1,$$

$$-k_2(X_1 - X_2) + k_4(X_2 - X_3) = m_2 \omega^2 X_2,$$

$$-k_3(X_1 - X_3) - k_4(X_2 - X_3) + k_5(X_3 - X_4) = m_3 \omega^2 X_3,$$

and

$$-k_5(X_3 - X_4) = m_4 \omega^2 X_4.$$

Thus

$$X_1[k_1 + k_2 + k_3 - m_1 \omega^2] + X_2[-k_2] + X_3[-k_3] + X_4[0] = 0,$$

$$X_1[-k_2] + X_2[k_2 + k_4 - m_2 \omega^2] + X_3[-k_4] + X_4[0] = 0.$$

$$X_1[-k_3] + X_2[-k_4] + X_3[k_3 + k_4 + k_5 - m_3 \omega^2] + X_4[-k_5] = 0,$$

and

$$X_1[0] + X_2[0] + X_3[-k_5] + X_4[k_5 - m_4 \omega^2] = 0.$$

Frequency equation is, therefore:

$$\begin{vmatrix} k_1 + k_2 + k_3 - m_1\omega^2 & -k_2 & -k_3 & 0 \\ -k_2 & k_2 + k_4 - m_2\omega^2 & -k_4 & 0 \\ -k_3 & -k_4 & k_3 + k_4 + k_5 - m_3\omega^2 & -k_5 \\ 0 & 0 & -k_5 & k_5 - m_4\omega^2 \end{vmatrix} = 0$$

45. Assume $x_1 > x_2 > x_3$. FBDs are then as follows.

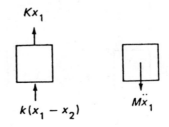

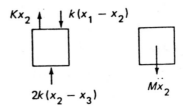

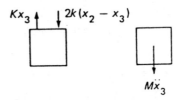

The equations of motion are therefore

$$-Kx_1 - k(x_1 - x_2) = M\ddot{x}_1,$$

$$k(x_1 - x_2) - Kx_2 - 2k(x_2 - x_3) = M\ddot{x}_2,$$

and

$$2k(x_2 - x_3) - Kx_3 = M\ddot{x}_3.$$

Substituting $x_i = X_i \sin \omega t$ and rearranging gives:

$$X_1[K + k - M\omega^2] + X_2[-k] + X_3[0] = 0,$$

$$X_1[-k] + X_2[K + 3k - M\omega^2] + X_3[-2k] = 0$$

and

$$X_1[0] + X_2[-2k] + X_3[K + 2k - M\omega^2] = 0.$$

The frequency equation is therefore

$$\begin{vmatrix} K + k - M\omega^2 & -k & 0 \\ -k & K + 3k - M\omega^2 & -2k \\ 0 & -2k & K + 2k - M\omega^2 \end{vmatrix} = 0;$$

that is,

$$(K + k - M\omega^2)[(K + 3k - M\omega^2)(K + 2k - M\omega^2) - 4k^2] + k[-k(K + 2k - M\omega^2)] = 0$$

or

$$M^3\omega^6 - \omega^4(3M^2K + 6M^2k) + \omega^2(3MK^2 + 12MKk + 6Mk^2) - (K^3 + 6K^2k + 6Kk^2) = 0.$$

The solutions to this equation give the natural frequencies.

46. Assume $x_1 > x_2 > x_3$. FBDs are

Equations of motion are therefore

$$-2k(x_1 - x_2) - k(x_1 - x_3) = 3m\ddot{x}_1,$$

$$2k(x_1 - x_2) = m\ddot{x}_2$$

and

$$k(x_1 - x_3) - kx_3 = m\ddot{x}_3$$

Putting $x_i = X_i \sin \omega t$ and rearranging gives

$$-3kX_1 + 2kX_2 + kX_3 = -3m\omega^2 X_1,$$

$$2kX_1 - 3kX_2 = -m\omega^2 X_2,$$

and

$$kX_1 - 2kX_3 = -m\omega^2 X_3;$$

that is

$$
\begin{bmatrix}
\dfrac{k}{m} & -\dfrac{2k}{3m} & -\dfrac{k}{3m} \\[2mm]
-\dfrac{2k}{m} & \dfrac{2k}{m} & 0 \\[2mm]
-\dfrac{k}{m} & 0 & \dfrac{2k}{m}
\end{bmatrix}
\begin{Bmatrix} X_1 \\ X_2 \\ X_3 \end{Bmatrix}
= \omega^2 \begin{Bmatrix} X_1 \\ X_2 \\ X_3 \end{Bmatrix}
$$

47.

$$
\begin{bmatrix}
\alpha_{11} & \alpha_{12} & \alpha_{13} \\
\alpha_{21} & \alpha_{22} & \alpha_{23} \\
\alpha_{31} & \alpha_{32} & \alpha_{33}
\end{bmatrix}
\begin{Bmatrix} X_1 \\ X_2 \\ X_3 \end{Bmatrix}
= \frac{1}{m\omega^2}
\begin{Bmatrix} X_1 \\ X_2 \\ X_3 \end{Bmatrix}
$$

Hence

$$
\begin{bmatrix}
0.5 & 0.25 & 0.25 \\
0.25 & 0.5 & 0.25 \\
0.25 & 0.25 & 0.5
\end{bmatrix}
\begin{Bmatrix} X_1 \\ X_2 \\ X_3 \end{Bmatrix}
= \frac{k}{m\omega^2}
\begin{Bmatrix} X_1 \\ X_2 \\ X_3 \end{Bmatrix}
$$

or

$$
\begin{bmatrix}
2 & 1 & 1 \\
1 & 2 & 1 \\
1 & 1 & 2
\end{bmatrix}
\begin{Bmatrix} X_1 \\ X_2 \\ X_3 \end{Bmatrix}
= \frac{4k}{m\omega^2}
\begin{Bmatrix} X_1 \\ X_2 \\ X_3 \end{Bmatrix}
$$

For lowest natural frequency assume mode shape 1, 1, 1:

$$
\begin{bmatrix}
2 & 1 & 1 \\
1 & 2 & 1 \\
1 & 1 & 2
\end{bmatrix}
\begin{Bmatrix} 1 \\ 1 \\ 1 \end{Bmatrix}
= \begin{Bmatrix} 4 \\ 4 \\ 4 \end{Bmatrix}
= 4 \begin{Bmatrix} 1 \\ 1 \\ 1 \end{Bmatrix}
$$

Hence correct assumption and

$$\frac{k}{m\omega^2} = 1,$$

so

$$f = \frac{1}{2\pi}\sqrt{\left(\frac{k}{m}\right)}\ \text{Hz.}$$

48. 775 kN/m; 3.52 Hz, 6.13 Hz. Unacceptable, $k = 1570$ kN/m.

49. 5.5%; 0.68%.

60. At one end, 10 dB and 16 dB; between 13 dB and 19 dB.

62. $Q = 14, 19.$

63. $\eta = 0.12.$

76. $$\left(\frac{\Omega_1}{\omega}\right)^2 + \left(\frac{\Omega_2}{\omega}\right)^2 = 2 + \mu,$$

and $\Omega_1\Omega_2 = \omega^2$.
   If $\Omega_1 = 250$, $\Omega_2 = 300^2/250 = 360$, and if $\Omega_2 = 350$, $\Omega_1 = 300^2/350 = 257$.
Therefore require $\Omega_1 = 250$ and $\Omega_2 = 360$ to satisfy the frequency range criterion.
(Ω_1 and Ω_2 are rev/min). Hence

$$\left(\frac{250}{300}\right)^2 + \left(\frac{360}{300}\right)^2 = 2 + \mu$$

and $\mu = 0.134$.
   Hence

   absorber mass $= 362$ kg.

and

   stiffness $= 142.9 \times 10^6$ N/m.

77. Substitute numerical values into frequency equation to give $m = 9.8$ kg.
   If $\Omega_1 = 85$, $\mu = 0.5$ so absorber mass $= 4.9$ kg, and $k = 773$ N/m.

78. $$\left(\frac{\Omega}{\omega}\right)^2 = \frac{2 + \mu}{2} \pm \sqrt{\left(\frac{\mu^2 + 4\mu}{4}\right)}.$$

If $\Omega_1 = 0.9\ \omega$, this gives $\mu = 0.0446$, and if $\Omega_2 = 1.1\ \omega$, $\mu = 0.0365$.
   Limit therefore $\mu = 0.0446$ and absorber mass is 134 kg with stiffness
30.1 kN/m.

79.

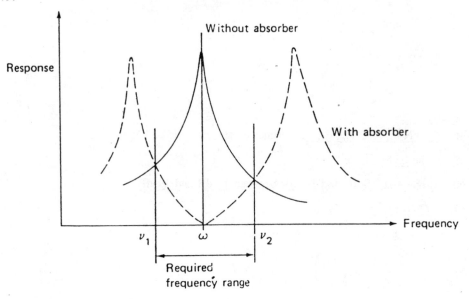

Require

$$\frac{F(K - mv^2)}{[(K + k) - Mv^2][k - mv^2] - k^2} = - \frac{F}{K - Mv^2} \quad \text{(phase requires } -\text{ve sign).}$$

Multiplying out and putting

$$\mu = \frac{m}{M} = 0.2$$

gives

$$2\left(\frac{v}{\omega}\right)^4 - \left(\frac{v}{\omega}\right)^2 (4 + \mu) + 2 = 0,$$

so

$$\left(\frac{v}{\omega}\right)^2 = \frac{4 + \mu}{4} \pm \tfrac{1}{4}\sqrt{(\mu^2 + 8\mu)} = 1.05 \pm 0.32.$$

Thus

$$\left(\frac{v}{\omega}\right) = 1.17 \quad \text{or} \quad 0.855,$$

so

$f_1 = 102$ Hz   and $f_2 = 140$ Hz.
Frequency range is therefore 102–140 Hz.

80. $\left(\dfrac{\Omega_1}{\omega}\right)^2 + \left(\dfrac{\Omega_2}{\omega}\right)^2 = 2 + \mu,$

and $\Omega_1\Omega_2 = \omega_1^2$

Now $\omega = 152$ Hz, $\Omega_1 = 140$ Hz so $\Omega_2 = 152^2/140 = 165$ Hz; hence

$\left(\dfrac{140}{152}\right)^2 + \left(\dfrac{165}{152}\right)^2 = 2 + \mu,$

and

$\mu = 0.0266.$

Require $\omega = 152$ Hz, $\Omega_1 = 120$ Hz   so $\Omega_2 = 192$ Hz (which meets frequency range criterion). Hence

$\left(\dfrac{120}{152}\right)^2 + \left(\dfrac{192}{152}\right)^2 = 2 + \mu^1$

so

$\mu^1 = 0.219.$

Therefore require $0.219/0.0266 = 8.2$, that is, 9 absorbers.

81. Cantilever absorber:

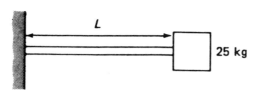

Beam stiffness at free end $= \dfrac{3EI}{L^3} = k.$

Thus

$k = \dfrac{3 \times 70 \times 10^9 \times (0.06)^4}{L^3 \times 12}$

Design based on 40 Hz frequency so

$$k = (2\pi \times 40)^2 \times 25.$$

Hence

$$\underline{L = 0.524 \text{ m.}}$$

When $f = 50$ Hz, calculation gives $\underline{L = 0.452 \text{ m.}}$

Bibliography

Beards, C. F., *Structural Vibration Analysis*, Ellis Horwood, 1983.

Beards, C. F., *Vibrations and Control System*, Ellis Horwood, 1988.

Beards, C. F., *Engineering Vibration Analysis with Application to Control Systems*, Edward Arnold, 1995.

Bickley, W. G. and Talbot, A., *Vibrating Systems*, Oxford University Press, 1961.

Bishop, R. E. D., Gladwell, G. M. L. and Michaelson, S., *The Matrix Analysis of Vibration*, Cambridge University Press, 1965.

Bishop, R. E. D., and Johnson, D. C., *The Mechanics of Vibration*, Cambridge University Press, 1960/1979.

Blevins, R. D., *Formulas for Natural Frequency and Mode Shape*, Van Nostrand, 1979.

Chesmond, C. J., *Basic Control System Technology*, Edward Arnold, 1990.

Close, C. M. and Frederick, D. K., *Modeling and Analysis of Dynamic Systems*, Houghton Mifflin, 1978.

Collar, A. R. and Simpson, A., *Matrices and Engineering Dynamics*, Ellis Horwood, 1987.

Crandall, S. H., *Random Vibration*, Technology Press and John Wiley, 1958.

Crandall, S. H. and Mark, W. D., *Random Vibration in Mechanical Systems*, Academic Press, 1963.

Davenport, W. B., *Probability and Random Processes*, McGraw-Hill, 1970.

Den Hartog, J. P., *Mechanical Vibrations*, McGraw-Hill, 1956.

Ewins, D. J., *Modal Analysis: Theory and Practice*, Research Studies Press, 1985.

Helstrom, C. W., *Probability and Stochastic Processes for Engineers*, Macmillan, 1984.

Huebner, K. H., *The Finite Element Method for Engineers*, Wiley, 1975.

Irons, B. and Ahmad, S., *Techniques of Finite Elements*, Ellis Horwood, 1980.

James, M. L., Smith, G. M., Wolford, J. C. and Whaley, P. W., *Vibration of Mechanical and Structural Systems*, Harper Row, 1989.

Lalanne, M., Berthier, P. and Der Hagopian, J., *Mechanical Vibrations for Engineers*, Wiley, 1983.

Lazan, B. J., *Damping of Materials and Members in Structural Mechanics*, Pergamon, 1968.

Meirovitch, L., *Elements of Vibration Analysis*, 2nd edn, McGraw-Hill, 1986.

Nashif, A. D., Jones, D. I. G. and Henderson, J. P., *Vibration Damping*, Wiley, 1985.

Newland, D. E., *An Introduction to Random Vibration and Spectral Analysis*, 2nd edn, Longman, 1984.

Newland, D. E., *Mechanical Vibration Analysis and Computation*, Longman, 1989.

Nigam, N. C., *Introduction to Random Vibrations*, Massachusetts Institute of Technology Press, 1983.

Piszek, K. and Niziol, J., *Random Vibrations of Mechanical Systems*, Ellis Horwood, 1986.

Power, H. M. and Simpson, R. J., *Introduction to Dynamics and Control*, McGraw-Hill, 1978.

Prentis, J. M. and Leckie, F. A., *Mechanical Vibrations; An Introduction to Matrix Methods*, Longman, 1963.

Rao, S. S., *Mechanical Vibrations*, Addison-Wesley, 1986, 2nd edn, 1990; *Solutions Manual*, 1990.

Richards, R. J., *An Introduction to Dynamics and Control*, Longman, 1979.

Robson, J. D., *An Introduction to Random Vibration*, Edinburgh University Press, 1963.

Schwarzenbach, J. and Gill, K. F., *System Modelling and Control*, 2nd edn, Arnold, 1984.

Smith, J. D., *Vibration Measurement and Analysis*, Butterworths, 1989.

Snowdon, J. C., *Vibration and Shock in Damped Mechanical Systems*, Wiley, 1968.

Steidel, R. F., *An Introduction to Mechanical Vibrations*, 3rd edn, Wiley, 1989.

Thomson, W. T., *Theory of Vibration with Applications*, 3rd edn, Unwin Hyman, 1989.

Timoshenko, S. P., Young, D. H. and Weaver, W., *Vibration Problems in Engineering*, 4th edn, Wiley, 1974.

Tse, F. S., Morse, I. E. and Hinkle, R. T., *Mechanical Vibrations, Theory and Applications*, 2nd edn, Allyn and Bacon, 1983; *Solutions Manual*, 1978.

Tuplin, W. A., *Torsional Vibration*, Pitman, 1966.

Walshaw, A. C., *Mechanical Vibrations with Applications*, Ellis Horwood, 1984.

Waterhouse, R. B., *Fretting Fatigue*, Applied Science Publishers, 1974.

Welbourn, D. B., *Essentials of Control Theory*, Edward Arnold, 1963.

Index